THE WRECK OF THE *PORTLAND*

THE WRECK OF THE *PORTLAND*

A DOOMED SHIP, A VIOLENT STORM, AND NEW ENGLAND'S WORST MARITIME DISASTER

J. NORTH CONWAY

An imprint of Globe Pequot, the trade division of
The Rowman & Littlefield Publishing Group, Inc.
4501 Forbes Blvd., Ste. 200
Lanham, MD 20706
www.rowman.com

Distributed by NATIONAL BOOK NETWORK

Paperback edition published in 2022. Originally published in hardcover by
Lyons Press in 2019.

British Library Cataloguing in Publication Information available

Library of Congress Cataloging-in-Publication Data

Names: Conway, J. North (Jack North), author.
Title: The wreck of the Portland : a doomed ship, a violent storm, and New England's worst maritime disaster / J. North Conway.
Description: Guilford, Connecticut : Lyons Press, [2019] | Includes bibliographical references and index.
Identifiers: LCCN 2019005687 (print) | LCCN 2019980556 (ebook) | ISBN 9781493039784 (cloth) | ISBN 9781493059461 (paper) | ISBN 9781493039791 (ebook)
Subjects: LCSH: Portland (Steamship) | Shipwrecks–New England–History–19th century. | Windstorms–New England–History–19th century. | New England–History–19th century.
Classification: LCC F9 .C67 2019 (print) | LCC F9 (ebook) | DDC 974/.03–dc23
LC record available at https://lccn.loc.gov/2019005687
LC ebook record available at https://lccn.loc.gov/2019980556

∞™ The paper used in this publication meets the minimum requirements of American National Standard for Information Sciences–Permanence of Paper for Printed Library Materials, ANSI/NISO Z39.48-1992.

Dedication

To all my granddaughters, Aria, Ella, and Harper.
Now you know what your Old Pappy does for a living, sort of.

Contents

"Dashed all to pieces! O, the cry did knock
Against my very heart! Poor souls, they perished!"

–William Shakespeare, *The Tempest*

Introduction

In his poem "Out, Out–" Robert Frost writes, "No more to build on there. / And they, since they / Were not the one dead, turned to their affairs," referencing a young boy who died in a chain saw accident. Those lines are as good as any to sum up the tragic events that befell the families of the passengers and crew who lost their lives when the ill-fated *Portland* sank on Thanksgiving weekend, 1898.

Although we as a society or individuals should not become hardened to death and tragedy, ultimately we must admit to ourselves and others, that life, our lives, must go on, regardless of the extent of a tragedy. The families of those who perished in the sinking of the *Portland*, among them Margaret Heuston, George Kenniston Sr., and Judith Cobb, all had to go on with their lives without their loved ones. They remembered and commemorated them, but despite their pain and suffering, they endured; they persevered. It was all they could do. The sea does not give up its dead or its many secrets.

There are many reasons why the *Portland* sank: the hubris of the ship's captain; a perfect storm raging up the New England coast; and a ship not built to withstand the open sea. Together they spelled doom for the ship, its crew, and its passengers.

Reflecting on this tragedy, it becomes important to remember not only those who were lost at sea but those who remained and their singular struggle to survive. These survivors are the real heroes and heroines in this sad, true story. Survivors always are.

BOOK ONE

The *Portland*

"the finest vessel that will travel the eastern waters."

–*The Portland Evening Express*, 1889

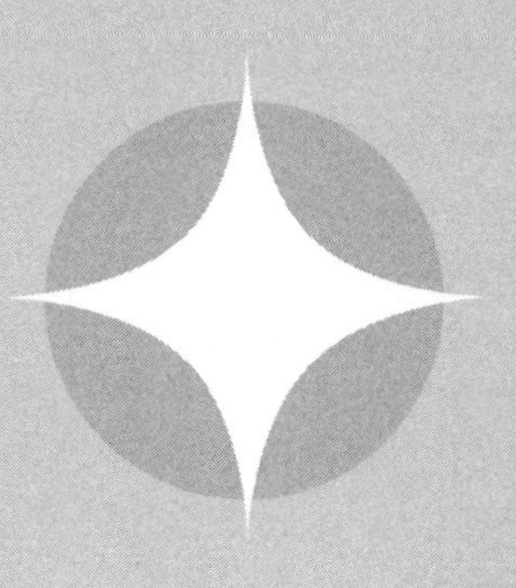

1: The Loom of Time

> I say so strange a dreaminess did there then reign all over the ship and all over the sea, only broken by the intermitting dull sound of the sword, that it seemed as if this were the Loom of Time, and I myself were a shuttle, mechanically weaving and weaving away at the Fates.
>
> –Herman Melville, *Moby Dick*, 1851

The fabric of this story is stitched together with the thread of Man and Nature; held fast in some places and frayed beyond repair in others. And like any story where Man challenges Nature over dominion of the land and sea, Nature wins out.

It was Saturday, the 26th of November, 1898, the end of the long Thanksgiving weekend, and hundreds of people were clamoring to take the last ferry back home from Boston to Portland, Maine. Among them was Hope Thomas, who had come to Boston to visit her sister and do her Christmas shopping. Employing a woman's prerogative, Hope made a crucial decision that forever sealed her fate. Emily Cobb, a talented songstress who had come to Boston for voice lessons had to return to Portland to perform at her first singing recital at the Portland First Parish Church on Sunday. George Kenniston Jr., a freshman English major at Bowdoin College, had to return home to resume classes at the prestigious school. There were many others.

On board the *Portland* was a skilled and seaworthy crew of sixty-eight men and women, many of them African Americans who lived in the Portland community of Munjoy Hill and were members of the Abyssinian Church. Among them was Chief Steward Francis Eben Heuston, who had spent thirty of his forty-four years working for the steamship company and was celebrating his first wedding anniversary that weekend.

Overseeing the crew and the passengers was fifty-three-year-old Captain Hollis Blanchard, whose bloodlines were rich in New England seafaring; his grandfather and father before him served as sea captains, and his reputation was that of a cautious and knowledgeable mariner.

And then there was the *Portland* itself, called "the finest vessel that will travel the eastern waters." Built in 1889, the ship had been making its regular run between Boston and Portland for ten years, carrying thousands of passengers and tons of cargo along the treacherous New England coast. Although one of the last of the side-wheel steamships, it had earned a reputation as a safe and seaworthy vessel.

Now, imagine if you will, a perfect winter storm forming off the New England coast that Saturday in November, at a time when there was no radio communication between ship and shore, no sonar to navigate by, and no sophisticated weather forecasting capacity. It was under these conditions that the *Portland* left Boston Harbor at 7 p.m. on the night of November 26 with two hundred passengers and crew on her regularly scheduled run back to Portland and ran headlong into a monstrous storm that is considered one of the worst winter gales of the nineteenth century. The storm destroyed hundreds of ships, homes, and businesses, and ended the lives of more than four hundred people along the New England coast.

The fates of the *Portland*'s crew and passengers were in the able hands of Captain Blanchard. Some contend it was hubris that drove Blanchard to leave the safety of Boston Harbor that day despite every indication that a severe New England storm was hurtling up the coast.

With all telephone and telegraph lines down in the aftermath of the storm, it was days before anyone was able to get word about the extent of the damages that had been done to life and limb. That included the whereabouts of the *Portland*. That is what this story is about–time and the tides that wait for no man, and the celestial weaving and weaving away at the Fates.

2: Red Sky at Morning

> When it is evening, you say, 'It will be fair weather; for the sky is red.'
> And in the morning, 'It will be stormy today, for the sky is red and
> threatening.' You know how to interpret the appearance of the sky,
> but you cannot interpret the signs of the times.
>
> –Matthew 16:2b–3

On Saturday, November 26, 1898, the luxurious steamship *Portland* was docked at India Wharf in Boston Harbor preparing for its regularly scheduled nine-hour, one-hundred-mile trip from Boston to Portland, Maine. The *Portland*, christened in October 1889, was the flagship of the Portland Steamship Company's Portland-to-Boston service. The 291-foot-long wooden side-paddle steamship could accommodate eight hundred passengers. Altogether there were 163 staterooms and more than two hundred passenger berths on the *Portland*'s upper and lower decks, fore and aft. It was described in brochures as a "floating palace."

The *Portland* had been crowded with passengers coming down from Maine to Boston to visit with family and friends for the Thanksgiving holiday on Thursday. It would be crowded on its return trip that Saturday evening, with passengers anxious to catch the last boat back home to Maine.

Captain Hollis Blanchard stood on the top deck of the ship, looking out across Boston Harbor. Dark storm clouds hung from the sky like tattered drapes being slowly drawn over the end of the day. The wind and temperature remained seasonable, with sporadic and brief snow flurries, not enough to even warrant the use of an umbrella. It was typical New England weather for late November.

Blanchard had checked with the Boston Weather Bureau earlier in the day. A fierce storm was reportedly barreling across the Midwest, while another storm was churning its way up the East Coast, the two storms headed on a collision course. Blanchard was not troubled by the report. He was certain he could stay well ahead of the storm if he left Boston on time at seven o'clock. Preparations were made to get underway that evening as planned.

3: Change in the Weather

On Friday evening, November 25, 1898, the Boston Weather Bureau had predicted a mild threat from the converging storm fronts. The initial forecast for Saturday and Sunday called for "fair continued cold, brisk westerly, shifting to southerly winds," for New England. For New York the forecast called for "increasing cloudiness and rain or snow by Saturday night." According to the forecast, on Sunday, the weather in New England would be "unsettled . . . rain or snow . . . possible, most likely in the afternoon or night. . . . The temperature [would] not change decidedly."

Saturday's weather forecast didn't change dramatically either, until later in the day when a special weather advisory was issued alerting maritime and other transportation concerns of higher than previously reported wind velocity and snow fall. According to this report, issued late on Saturday, November 26, Maine, New Hampshire, and Vermont could expect "heavy snow" on Saturday and temperatures that were "much colder; southeasterly winds shifting by tonight to northeasterly gales." For Massachusetts, Rhode Island, and Connecticut, heavy snow was predicted, with clearing on Sunday followed by much colder weather. According to the advisory, there would be, "northeasterly gales tonight, and northwesterly gales by Sunday."

The first weather forecasting bureau was established in 1814 as part of the United States Medical Department. But weather forecasting for seamen like Blanchard was neither accurate nor trustworthy. The weather forecasting technology was unsophisticated and the reporting haphazard.

In 1870, a national weather service was created by Congress during the Grant administration. President Ulysses S. Grant signed into law a joint resolution with Congress that empowered the secretary of war to establish a national weather service. The act called for the secretary to establish this service for the purposes of, "taking meteorological observations at the military stations in the interior of the continent and at other points in the States . . . and for giving notice on the northern (Great) Lakes and on the seacoast by magnetic telegraph and marine signals, of the approach and force of storms."

The weather service was placed under the secretary of war's supervision because it was felt that in doing so the greatest degree of accuracy and regularity would be accomplished. Oversight for the new service was placed within the Signal Service Corps under the leadership of its commander, Brigadier General Albert J. Myer. Myer designated the new weather service as The Division of Telegrams and Reports for the Benefit of Commerce. Weather observations were assigned to members of the Army Signal Service at more than twenty observing stations throughout the country, and weather forecasts were regularly sent to the Washington, D.C., headquarters on a daily basis.

Weather forecasts were made for eight regions of the country on a basis of twenty-four hours in advance and included predictions based on four elements: weather, wind, pressure, and temperature. In 1886, forecasts were focused on individual states rather than the larger districts made up of several states. Two years later, in 1888, weather forecasts were extended from their previous twenty-four-hour durations to thirty-six hours.

By 1898, forecasts were extended to cover a forty-eight-hour period. The number of weather forecasting field offices expanded from a mere 24 offices throughout the country to 284 offices by 1878, and weather observations were recorded and sent to the Washington, D.C., headquarters three times a day. From headquarters, forecasts were distributed to steamship companies, railroads, farm bureaus, observers, and various newspapers. The forecasts, compiled from the various daily telegraphed reports from throughout the country included information on the changes in barometric pressure, temperature, humidity, wind velocity and cloud coverings.

Because meteorology was relatively embryonic during the late 1800s, forecasts didn't always prove to be accurate; however, it was noted by weather bureau officials that "while scientists cannot tell at what hours to carry an umbrella, they can tell when great storms and waves of intense heat or cold are coming so as to be of great value to all the industries of the land." But weather bureaus were limited in their abilities to accurately predict weather conditions, since they were unable to measure the changing patterns that percolated and formed high above Earth's atmosphere.

4: Proverbs and Predictions

Many unscientific maritime observers still relied on weather forecasting idioms and proverbs to predict the weather. Among them were: "a red sun has water in his eye"; "clear moon, frost soon." Other popular forecasting ideas were, for example, that much noise made by rats and mice indicates rain; anvil-shaped clouds are very likely to be followed by a gale of wind; if rain falls during an east wind, it will continue a full day; a light yellow sky at sunset means wind; a pale yellow sky at sunset means rain. For seafarers, the most common of all was, "Red sky in morning, sailors take warning; red sky at night, sailors' delight."

This old mariners' proverb had its origin in Matthew 16:2b–3, when Jesus spoke to the fishermen, telling them that they knew how to predict the weather but had no understanding of the times they lived in. Thousands of years later, through the advances in technology, the Library of Congress would lend credence to Mathew's proverb:

> A red sunrise reflects the dust particles of a system that has just passed from the west. This indicates that a storm system may be moving to the east. If the morning sky is a deep fiery red, it means a high water content in the atmosphere. So, rain is on its way. (Library of Congress, "Weather Facts," 1949)

5: Bloodlines

Captain Blanchard had met with John W. Smith, the chief meteorologist at the Boston Weather Bureau office, to go over the most recent weather forecasts. He learned from Smith that the storm bearing down from the Great Lakes region was making its way toward the New England coast on a path that would intersect with a second storm that was moving up the coast from the South. Based on this information, a special weather advisory was issued by the Weather Bureau's Washington, D.C., headquarters calling for heavy snowfall and gale winds all along the New England coast beginning on Saturday evening and lasting into late Sunday. Blanchard took it all into consideration. Everyone knew him as a cautious man and a thoughtful captain.

The fifty-three-year-old Blanchard was tall and lean, always immaculately dressed in a dark blue double-breasted top coat and gold braided captain's hat that covered his thick black hair and rested above his slightly protruding ears. His chestnut eyes were resolute and knowing. His moustache was neatly trimmed and combed, waxed into place, extending down both corners of his mouth and ending in two fine points. His thick beard, black except for flecks of gray along his chin, was meticulously groomed; not a hair was out of place. Except for the Bible, nautical maps, and weather forecasts, Blanchard reportedly read nothing–not even the daily newspapers.

His bloodlines were rich in New England seafaring. He was born in Searsport, Maine, in 1845 and spent many years living just down the coast in Belfast. He lived now in Westbrook with his wife, two sons, and a nineteen-year-old daughter. His father and grandfather before him had been sea captains, as were his two brothers. His formal education was limited except for what he'd learned from the sea and other seamen.

He went to sea at nine years old as a cabin boy on one of his father's vessels and rose rapidly through the ranks to pilot, a position where he was responsible for navigating the ship under the instructions of the captain. He began his merchant marine career in the 1880s. His first job was on the schooner, *Mary B. Coombs* out of Isleboro, Maine. Bath, Maine, had a more than a dozen shipyards building schooners of all sizes during most of the nineteenth century. Blanchard later went on to serve on the brig *R.S. Hassell* and the steamships *Cambridge* and *William Tibbetts*. The brig, a sailing ship with two generic, square-rigged masts, was fast and maneuverable and used as a merchant vessel. However, they fell out of use with the advent of steamships because they were difficult to sail into the wind and because they required a relatively large crew for their otherwise small size.

He began working for the Portland Steamship Company nearly a decade previously. For nine years he diligently worked his way up from mate to pilot. He had served as pilot on the *Bay State* as well as the *Portland*. Although he had served as captain on board other vessels, he had only recently been given full command of the *Portland* three weeks ago, following the death of Captain William Snow, the former longtime captain of the ship. Following Snow's death, Blanchard was promoted to captain and put in charge of the steamship. Although an experienced sailor, this was his first command within the Portland Steamship Company. According to Harry Gratwick, in his book, *Historic Shipwrecks of Penobscot Bay*, Blanchard was known as "an honorable man, a good navigator and master of his calling."

6: Churning up the Coast

There were conflicting reports about the pending storm. According to the Weather Bureau's Washington, D.C., home office, a high-pressure system had settled across Ohio and the Midwest, enveloping that entire section of the country in a relatively seasonal warm front. Chillier winds were blowing in from the Great Lakes region followed by what most observers considered trifling snow flurries, brisk gusts of wind, and not terribly noticeable showers. There was no cause for alarm at the Weather Bureau's main offices.

According to the Boston Weather Bureau, however, the storm front moving across the Great Lakes region would pick up intensity by Saturday morning, heading straight for the East Coast and bringing with it a blast of arctic temperatures and plummeting snow. The Midwestern storm front was churning its way northeast, gaining wind speed and momentum, reportedly on a collision course with the storm moving up the eastern seaboard from the Gulf of Mexico.

According to a Boston weather advisory, issued late on Saturday, November 26, Maine, New Hampshire, and Vermont could expect "heavy snow" on Saturday and "snow and much colder; southeasterly winds shifting by tonight to northeasterly gales." For Massachusetts, Rhode Island, and Connecticut, heavy snow was predicted, with clearing on Sunday followed by much colder weather. According to the report, there would be "northeasterly gales tonight, and northwesterly gales by Sunday." No one would have been able to accurately predict the pending storm's ultimate monstrous fury.

7: Eben Heuston

The *Portland*'s chief steward, Francis Eben Heuston, had spent thirty of his forty-four years working for the steamship company, beginning first as a lowly cabin boy and working his way up to the lofty position he now held. It was a special trip back home to Portland that weekend for Heuston. Tucked in the pocket of his starched white vest was a gold cross that he'd been secretly saving up to buy for his wife, Margaret Ann. He'd bought it at one of the finest jewelry stores in downtown Boston. This weekend was their anniversary. They had been married one year to the day on that Saturday.

8: Emily Cobb

Twenty-three-year-old Emily Cobb moved with reckless abandon through the busy holiday streets of Boston, the hood of her dark wool cape pulled up over her head. She raced in and out of crowds of leisurely shoppers heading toward India Wharf carrying her sheet music and hymnal in the bag flung over her shoulder. She could not afford to miss the last boat back to Maine. On Sunday, she was scheduled to give her first solo singing recital at the First Parish Church in Portland. At the suggestion of First Parish's associate pastor, John Carroll Perkins, she had come to Boston on the busy weekend to take singing lessons in preparation for the recital. Perkins had paid for the trip and was certain it was money well spent, since Emily had a stunning voice.

9: A Woman's Prerogative

Every year, Hope Thomas came down from Portland to Boston and stayed with her younger sister Grace and her husband. The day after Thanksgiving was the biggest shopping day of the year. The stores were all decorated for Christmas and the stores filled with shoppers drawn in by the holiday sales.

Hope had already bought all her Christmas presents. They were wrapped and bundled and she carried them in two large shopping bags that she held in each hand as she awkwardly maneuvered through the downtown Boston crowds. Everybody was coming and going, hustling and bustling through the streets.

Besides all the presents for her husband and family, she had purchased a pair of shoes for herself at a downtown store but was unable to make up her mind about keeping them. It would have been a long trip back to return them, and she was running out of time. The *Portland* would be disembarking soon, and she had promised her husband, Bill, she would be coming home on the last boat on Saturday. Bill Thomas was the captain of the fishing schooner, *Maud S.*, out of Bailey's Island. He had offered to bring her to Boston on board the *Maud S.*, but Hope insisted on taking the steamship. She didn't want the stench of fish all over her clothes when she arrived in Boston to visit with her sister. That was her prerogative, she maintained, just as it was her prerogative to change her mind about the shoes she had bought if she wanted to, and heading toward the pier to purchase her ticket home, she still hadn't made up her mind. Bill Thomas was out at sea that weekend aboard the *Maud S.* Hope had heard the rumors of a pending storm, but she knew Bill would be safe and would put up somewhere along the coast if a storm did hit. She imagined she would make it back home to Portland before her husband docked.

10: Do You Know What You're Doing?

Despite the troubling weather reports, Blanchard had already made up his mind. He spoke with another Maine sea captain, C. H. Leighton, about the predicted storm. Leighton, from Rockland, Maine, had intended taking the *Portland* back home that evening but subsequently changed his mind when he heard the news about the storm. Leighton asked Blanchard about his decision to leave Boston that evening in the face of the ominous weather forecast. He advised Blanchard against going, but it appeared Blanchard, despite all that was known and told to him, had already made up his mind to go.

LEIGHTON.
You think you'll be going?
BLANCHARD.
I think I shall.
LEIGHTON.
I don't think this is a fit night to leave port.
BLANCHARD.
I don't know about that. We may have a good chance.
LEIGHTON.
Do you know what you're doing?
BLANCHARD.
I do.

Considering the number of lives Blanchard held in his hands, having a "good chance" did not seem like a reasonable response to a frustrated Leighton.

11: Ritual and Routine

"I must go down to the seas again, to the lonely sea and
the sky,
And all I ask is a tall ship and a star to steer her by."

–John Masefield, "Sea Fever," 1902

Captain Hollis Blanchard was a creature of ritual and routine–the ritual of command and the routine of maritime life. He served only two masters; God and the Portland Steamship Company, and should there ever be a conflict between the two, Blanchard would always come down on the side of the company. Although God had His infinite dominion over land and sea and all those who inhabited them, the steamship company had province over the Boston–Portland ferry and Blanchard considered himself the divining instrument of this earthly compact.

The ritual of his command on board ship included all things both great and small. Only three months previously, he had been appointed captain of the *Portland.* Blanchard required all those serving under him to address him as Captain Blanchard. He, in turn, always addressed his crew by their title and last names. It was always, "First Pilot Strout," or "Chief Steward Heuston," unlike his First Pilot, Lewis Strout, who knew all the crew and officers and called them by their first names. But Strout was one of the new breed of steamship officers, who didn't follow Blanchard's path through the ranks to his position. Strout had graduated from the Massachusetts Nautical Training School established in 1891 and located along India Wharf. The school later moved to Cape Cod in 1936 and then changed its name to the Massachusetts Maritime Academy in 1942. It remains there to this day and is the second-oldest maritime academy in the country, the first being the SUNY Maritime College located in New York, founded in 1874. The ranks of the steamship company were being populated with younger academy-trained seamen like Strout and Alexander Dennison, who was recently given command of the *Bay State*, the newest steamship in the company's fleet.

Blanchard never appeared on board without being attired in his full uniform and captain's hat, and he expected his crew to do the same. He would severely reprimand anyone in his crew who was out of uniform, whether sailor or a steward, man or woman. There were two women on board the *Portland* crew, stewardesses, Marjorie Berry and Carrie Harris, both from the African American community of the Munjoy Hill section of Portland.

Blanchard was exceptionally proud of his crew and the ship's appearance, reputation, and record. He was a stickler when it came to detail, making sure the stiff, white, high collar on his shirt appeared no more than two inches above the navy blue spread collar of his double-breasted, gold buttoned dress uniform.

His routine was always the same. He saw to every detail on the ship, from checking the amount of coal on board used to fire the steamship's great engine to making sure the ship's ornate saloon was adequately stocked. Blanchard, however, was a teetotaler. Although the *Portland* had a well-stocked bar of rum, wine, ale, and brandy, it was there for passengers only if they chose to imbibe. But he took note of those under his command whom he deemed too fond of the drink. Without the slightest compunction he had fired men and women alike whom he suspected of drinking while on duty. He once fired a lowly porter because he suspected the man of pilfering rum from the ship's saloon below deck. Blanchard claimed he could smell the liquor on the man from across an entire room. The man protested, claiming that what the captain smelled on him was accidentally spilled on his jacket by the bartender. But although the bartender confessed to spilling the drink on the cuff of the porter's jacket, Blanchard remained steadfast in his judgment. There was nothing anyone could do to convince him otherwise, and the porter was fired on the spot. Blanchard never backed down, even if he was wrong.

He checked the provisions in the ship's kitchen, making sure there was a good supply of steak, pork, salt cod cakes, beans, mashed potatoes, chicken, prime rib, and especially this weekend for the trip back to Maine, carved turkey breast and cranberry sauce. He randomly inspected cabins to make sure clean linens had been provided and checked that all lamps along the deck were lit and filled with oil and that every railing, inside and out, had been cleaned and polished. He had crew who were assigned to do the same thing, and they dutifully reported back to him, but Blanchard treated the *Portland* as if it were his own private vessel and each passenger was a private guest of his. He checked and rechecked every detail on board the ship routinely before it disembarked. This weekend was no different.

"Discipline, Knowledge, Leadership"

–Motto of the Massachusetts Maritime Academy,
Buzzards Bay, Massachusetts

12: Officers and Crew

Blanchard had a crew of sixty-eight men and women including the ship's pilots, engineers, oilers, firemen, deckhands, porters, stewards, stewardesses, cooks, watchmen, and saloon keepers. Many were African Americans who lived in the Portland community of Munjoy Hill and were members of the Abyssinian Church. Among the officers aboard the *Portland* was First Pilot Lewis Strout. As first pilot, the Massachusetts Nautical Training School graduate Strout was responsible for the navigation of the ship under the captain's directions. The second pilot was Lewis Nelson.

The chief engineer was Thomas Merrill. He was responsible for maintaining the ship's engines and boilers and all the operations and maintenance that had to do with any and all engineering equipment throughout the entire ship. His staff included Second Engineer John T. Walton, First Assistant Engineer John Walton, and Second Assistant Engineer Charles Merrill.

James McNeil and John Dillon served as the ship's oilers. Armed with oil cans, grease guns, rags, and brushes, their job was to keep the ship's machinery well oiled and working.

Harry Rollinson was one of the ship's firemen or stokers. Rollinson and the other engine stokers who labored in the bowels of the ship fired the furnace with coal and kept the furnaces constantly fired. While one set of stokers furiously shoveled coal into the roaring fire and flames, others shoved their slice bars, known as lances, deep inside the open doors of the huge furnaces and, using brute strength, heaved the bank of fire up several inches. They then yanked the lances out so that the opening in the fiery coal stack sucked in a draft of air that ignited the flames into a furious roar.

Wearing thick gloves to protect their hands from the flames and heat, stokers like Rollinson labored in the engine room shirtless, their heavy canvas trousers blackened with soot and coal dust. Rollinson was a big man with broad shoulders and massive arms. In many places along his forearms and even his chest, his skin was scarred with wounds from burns.

For serving the passengers, the ship had a staff of porters, stewardesses, saloon keepers, bartenders, cabin boys, cooks, and stewards like Chief Steward Eben Heuston. As chief steward, Heuston was in charge of overseeing all the staterooms and cabins on board the *Portland*, as well as the dining room, storerooms, and kitchen. All the soups, fish, meats, and vegetables were prepared and cooked in one room, and the bread and pastry in another. Next to the kitchen was the serving room, or pantry. Polished silver coffee pots, cups, fine china, crockery, silverware, saucepans, and serving urns and platters were stored there.

The ship's purser was Frederick Ingraham. Ingraham was responsible for the handling of all the money on board, as well as maintaining the ship's supplies and cargo, including the ship's coal. There were more than sixty-five tons of coal aboard the *Portland,* although the usual nine-hour trip from Boston to Portland used only twenty-five tons. On the wharf side of the ship, the coal was brought on board in wheelbarrows up a gangway and stored in bunkers below deck in the engine room. Ingraham was also responsible for the passenger manifest. The only manifest listing all the crew and passengers on the *Portland* was in Ingraham's safekeeping on board the ship. No manifest was required to be left on shore.

13: Charming and Courteous

Chief Steward Eben Heuston was more than adept at handling all types of passengers, from the difficult to the frightened. He acted like a mother hen to all of them, coaxing, cajoling, acquiescing–and, when needed, being solicitous–whatever it took to make the voyage pleasant and uneventful for everyone on board. Heuston was a charming, courteous man who was seemingly everywhere on the *Portland* during its voyage. He was always immaculately dressed in ironed dark trousers, white waistcoat, and white gloves. He was a handsome man with coffee-colored skin, curly white hair, and a dark, trimmed moustache. Meticulous and courteous to everyone, Heuston was the best at what he did.

Heuston had married Margaret Ann Ball in a small private ceremony held at the Abyssinian Church a year before. They were celebrating their first anniversary that weekend. It was the first marriage for Heuston but the second for his wife, who knew all too well the unrelenting fury of the unforgiving sea. She had lost her first husband, William Ball, in 1892 when a fishing boat went down in Casco Bay, leaving her a widow with two young children to care for.

She was born in Brunswick, Maine, in 1863 and had spent her entire life in Portland, where she and Eben now owned a home on Munjoy Hill, where most of Maine's African American community had settled. The Hill was a tight-knit community. The houses there were neat, clapboard, well-cared-for, single- and two-story structures built side by side, all identically whitewashed. They occupied both sides of a narrow cobblestone street running along the slope of a hill leading down to Portland Harbor.

Both Eben Heuston and his wife were active members of the Abyssinian Church. The three-story, whitewashed clapboard church with stained-glass windows and pointed brass-covered steeple was located on Newbury Street at the top of the hill, overlooking the entire community. Twenty of the crew on board the *Portland* came from Munjoy Hill and were members of the Abyssinian Congregational Church. Established in 1828, the church was Maine's oldest African American church building, and the third-oldest in the nation. Established in order to meet the demand for a place to worship for African Americans in Portland, it was also the cultural center for African Americans in southern Maine. Besides being a place for worship and revivals, it was the organizing site and meeting place for abolition and temperance meetings, lectures and concerts, the Female Benevolent Society, the Portland Union Anti-Slavery Society, and Negro conventions. The church had served as a hiding place for runaway slaves, as part of the northern "underground railroad" network during the Civil War. It also ran one of the first African American schools in Portland, from the mid-1840s through the mid-1850s. Margaret Heuston taught Sunday school at the church.

The church was nicknamed "The Anchor of the Soul" because so many members of the congregation, like Eben Heuston, were mariners. Heuston couldn't wait to get back home to give his wife the gold cross he had bought for her to celebrate their one-year anniversary. Despite rumors of a winter storm on the way, Heuston hoped that Captain Blanchard would make the regularly scheduled run back to Portland so he could celebrate his anniversary with his wife at home.

14: Safety Record of the Portland

The *Portland* had been making its regular run between the two cities for the past ten years, carrying thousands of passengers and tons of cargo along the treacherous New England coast and it had earned an admirable reputation as a safe, dependable, and seaworthy vessel. It never had an accident and had never missed leaving or arriving in port on time.

Passengers used the steamships for a variety of reasons, both business and pleasure. In the summer months, the steamships were often crowded with vacationers escaping the heat. Portland was the gateway to Maine's summer resorts, and tourists packed the *Portland*'s cabins en route to summer relaxation. Passengers also utilized the *Portland* for business travel. While Boston was already an established center of trade by the 1890s, drawing businessmen, lawyers, and manufacturers to its hub, Portland was growing at a fast pace and becoming Maine's coastal center of manufacturing, drawing investors up from Boston. The *Portland*'s timely, regular service and comfortable cabins made for relaxing efficient transportation along the coast. In the fall and winter, passengers traveled from Maine to Boston to visit relatives during the holidays and to shop in downtown Boston for Christmas presents.

The company took advantage of its nearly pristine safety record in its advertising. An 1897 brochure for the steamship line boasted, "After fifty-two years of service to the traveling public, this Company, with new and powerful steamers, unrivalled passenger accommodations, careful and experienced officers in every department, will endeavor to give the same care through which, in more than half a century no passengers have lost life or received injury to person or property." The *Portland* was equipped with eight metal life boats, four metal life rafts, and more than eight hundred cork-filled life jackets.

By 1897, the company was running three steamships between the two cities: the *Bay State*, the *Portland*, and the *Tremont*. The *Portland* was more luxurious than the other two ships, but the *Bay State* was a newer and faster ship. It had been built several years after the *Portland*. Instead of a single paddle wheel in the stern of the ship, they both were constructed with immense paddle wheels on each side. The *Portland* and the *Bay State* were the last of the side-wheel steamships.

The Portland Steamship Company had become the premiere steamship service between the two New England ports due in no small part to its record of prompt, safe passage and its luxurious vessels. Still, there were many who advised that although beautiful and luxurious, the side-paddle steamers were unsafe on stormy seas and treacherous currents. This, however, did not seem to dissuade passengers from taking the low-cost vessel. A round-trip ticket from Boston to Portland cost the modest sum of $2.

15: A Conversation with Dennison

Although Blanchard had spoken briefly with fellow Maine sea captain C. H. Leighton about the predicted storm, he wasn't the only person he spoke with. Blanchard discussed the situation with Alexander Dennison, the newly appointed captain of the *Portland*'s sister ship, the *Bay State*, which was still docked in the safety of Portland Harbor. Dennison telephoned Blanchard in Boston. The thirty-year-old Dennison was another product of the Massachusetts Nautical

Training School. Blanchard referred to him as the "kid-pilot" because of his relatively young age.

The two men discussed the weather forecast. Dennison informed Blanchard that the weather in Portland, "looked black." He suggested to Blanchard that he might consider waiting until later that evening to see how the storm front turned out before leaving Boston. He told him that John Liscomb, the company's new manager, suggested holding the *Portland*'s departure until 9 p.m. to see how the storm developed off the coast. Blanchard responded curtly to Dennison's suggestion. He told Dennison that if he wanted to spend Sunday in Portland that was up to him but that he had enough seafaring experience to know whether he could beat the storm or not heading out of Boston.

It may well have been Dennison's telephone call that sealed the fate of the *Portland*'s crew and passengers. Disappointed that the younger and less experienced Dennison had been given command of the newer *Bay State* steamship over him, Blanchard may have been trying to prove his worth and skill to Dennison and the company and, more so, establishing that his nautical prowess far exceeded the new young captain. According to a *Boston Evening Transcript* article published many years later, it was alleged that it may have been Blanchard's disappointment at not receiving the promotion to captain of the *Bay State* that spurred Blanchard on to leave Boston Harbor that fateful evening.

> The fact that the two captains talked over the phone that day has given rise to the to the popular legend that Captain Blanchard sailed contrary to the advice of the general manager because he was anxious to prove his sailing abilities superior to those of Captain Dennison by steaming into Portland Harbor where the *Bay State* would still be at the wharf." (Edward Rowe, *Snow, Storms and Shipwrecks of New England*, 1943)

When the former captain of the *Bay State* suddenly passed away, it created a position for a new captain. Blanchard had anticipated being promoted to it. The *Bay State* was the newest steamship in the fleet and Blanchard had the most seniority of any captains. The Portland Steamship Company had also recently lost its general manager, John B. Coyle Jr., who passed away in late November. He was replaced by former Boston-based company agent, John F. Liscomb.

Blanchard had anxiously waited for the telephone call from the steamship company offices in Portland for news of his new promotion, but the telephone call never came. Instead, the younger, less experienced, Alexander Dennison, was given command of the *Bay State* over him. Blanchard was both surprised and disappointed when he learned that he had been passed over. He brooded over it.

Giving Dennison the appointment as captain of the *Bay State* was a mistake, Blanchard thought. Maybe the inexperienced Captain Dennison might find the predicted storm too difficult to navigate, but he would not. Blanchard figured he'd be in port well ahead of the storm, while the inexperienced Dennison clung to safe harbor in Portland.

16: The Birth of the Portland Steamship Company

Hollis Blanchard was only a year old when sea captain Joseph Brooks of Kennebunkport, Maine, founded the Portland Steam Packet Company with a group of partners in 1844. It was incorporated by the Maine legislature in 1845. Brooks began the company by operating two propeller-steam freighters running in opposite directions between Boston and Portland.

Storms were the greatest financial challenge he had to face. His insurance burden cut deeply into his profits. Within a few years, commodious new side-wheel passenger steamships with cabins, finished in cherry and mahogany, were added to the line. They ran at night between Franklin Wharf in Portland and India Wharf in Boston and coincided with railroad schedules at either end.

The company's first ship, the *Commodore Preble*, made its first run from Portland to Boston in 1844. Despite competition from the railroads running regular service between the two cities and other steamship service, the Portland Steam Packet carried more than twenty-five thousand passengers and nearly $50,000 in freight during 1848, four years after its inauguration. The company later became the Portland Steamship Company.

During the course of thirty-seven years under the management of Brooks, the Portland Steamship Packet Company transported millions of passengers and not one was ever hurt or lost. The line had the best safety record by far, and they carried much less insurance than any of the other companies and retained a higher percentage of their fares. The Portland Steam Packet Company provided continuous service between Boston and Portland from 1844 until 1901.

Chapter 271.

AN ACT to incorporate the Portland Steam Packet Company.

Be it enacted by the Senate and House of Representatives in Legislature assembled as follows:

SECT 1. Lemuel Dyer, William Kimball, Daniel Winslow, Samuel Clark, Nathaniel Blanchard, and Moses Gould, their associates successors and assigns are hereby created a corporation by the name of the Portland Steam Packet Company for the purpose of steam navigation between Portland, Boston and such other ports as may be deemed useful, with all the powers and privileges, and subject to all the duties and liabilities that may be granted and required concerning similar corporations, by the laws of this state.

SECT 2. The said corporation may take, purchase and hold real and personal estate, to an amount not exceeding at any one time, one hundred thousand dollars, with full power to manage and dispose of the same.

SECT 3. Any two persons named in this act, may call the first meeting of said corporation, for the purpose of organizing, at such time and place as they may see fit by giving notice thereof two weeks successively, in some newspaper published in Portland.

Approved March 31, 1845. (Twenty-Fifth Legislature of the State of Maine, 1845)

17: Old Probabilities

Joseph Brooks had earned his nickname "Old Probabilities" based on his pioneering endeavors in the field of weather forecasting. Born in Auburn, Maine, in 1806, he became intrigued in the process of weather prediction and its potential use in the seafaring trade in 1841 after he attended a lecture given by noted professor and meteorologist James P. Espy. Espy theorized that storms moved eastward across the country and that a storm reported in New York could be projected to hit the Maine coast within a matter of one or two days. According to Espy, because of the many advances in telegraphy, it was possible for weather reports generated in places like New York and Philadelphia to be sent in a timely manner to judicious mariners, like Brooks, to use and base their sailing schedules on. He published his theory of storms in 1841 in his book *The Philosophy of Storms*. Espy became the chief meteorologist for the United States War Department in 1842 and the Department of the Navy in 1848. He also served as the chief meteorologist for the American Philosophical Society in Philadelphia, where he was instrumental in establishing a network of weather observers to track and study storms. He was able to persuade the Pennsylvania legislature to arm an observer in each county with a barometer, thermometers, and rain gauges.

Brooks became a believer in Espy's theory and a supporter of weather forecasting. According to the Portsmouth, New Hampshire, online publication, Seacoastonline.com, "Before 1850, against the better judgment of his business partners, Brooks had employed agents in New York, New Haven, Springfield, Boston and Portland to make observations of the state of the wind and weather and to send their findings to him every day over telegraph wires. If the weather looked bad in the morning up to three additional reports were made each day."

His persistence in the face of skepticism, even among his co-owners at the Portland Steamship Packet Company, helped preserve his company's profits and saved lives. In February 1852, Brooks was involved in a test case that proved him right to his otherwise skeptical partners in the company. According to Brooks, "I sent a telegram (telegrams on this subject passed daily between the Boston and Portland offices of the company) to the agent in Portland at 12 o'clock noon, to the effect that a heavy snow storm was raging in New York but that the weather continued fine in Boston. At four o'clock in the afternoon another telegram was sent, stating that the storm had reached Springfield, and the Boston boat would not leave her dock and that if the *St. Lawrence* (then a new boat) left Portland, she would find herself in the midst of the storm before the passage was half completed. Now sneers and jeers were in order. The Portland agent came to the conclusion that storms in New York had nothing whatever to do with the weather in Boston and Portland or in between those points and sent his ship to sea."

The *St. Lawrence* left Portland with 307 passengers on board. It ran headlong into the storm that had been predicted off Portsmouth, New Hampshire. The storm ravaged the ship, and by the time it reached Boston Harbor it was nearly inoperable. The *St. Lawrence* remained adrift just outside the harbor for three days after her rudder was broken in the storm. Although most of her cargo was lost, fortunately all her passengers were saved. The near disaster, despite Brooks's warning and the various weather predictions proved to his partners that "Old Probabilities," was right, and weather forecasting became a vital part of the company's overall operations. By the time Brooks retired from the company in 1881, his system of using weather forecasting and observations to reduce losses had been widely adopted by most steamship companies.

Because of James Brooks's championing of weather forecasting as a vigorous part of the steamship company's operations and his willingness to bend to the will of Nature when it came to storms, the day's most sophisticated weather reports were available to Blanchard, predicting what could be a major storm along the east coast. It is ironic, then, that Blanchard, heir to Brooks's visionary efforts, took no heed of them.

In another irony of the day, one of the partners with Brooks in the steamship company was Captain George Knight, whose daughter, Judith, married Portland organist and musician Edward Cobb. He and Judith were the parents of Emily Cobb, who was scheduled to sail out of Boston Harbor on the night of November 26 on board the *Portland*.

18: For the Beauty of the Earth

Emily Cobb planned to begin her solo performance at the First Parish Church in Portland on Sunday with Sandford Pierpoint's hymn, "For the Beauty of the Earth." It had been a favorite of her father's.

Pierpoint (1835–1917) had written the hymn while taking a walk through the woods at his home in Bath, England. Overwhelmed with the beauty he witnessed around him, Pierpoint sat down and wrote it. He told friends that he had written the hymn to thank God for the wondrous beauty he had created, but also to thank God for the gift of his friends and family.

> "For the glory of the skies,
> For the love which from our birth
> Over and around us lies.
> Lord of all, to Thee we raise,
> This our hymn of grateful praise.
> For the beauty of each hour,
> Of the day and of the night,
> Hill and vale, and tree and flower,
> Sun and moon, and stars of light."
>
> (Sandford Pierpoint, "For the Beauty of the Earth," 1864)

Cobb's trip to Boston on her own was the first time she had been away from her home in Portland. She left her clerk's job at Owen, Moore & Co.'s retail store in Portland to pursue a full-time career in music, following in her late father's footsteps. Edward F. Cobb had been one of the most celebrated musicians in the City of Portland prior to his death.

Edward F. Cobb had been a renowned pianist, organist, singer, and music teacher. He died of tuberculosis when Emily was sixteen years old. She and her father had a special bond between them, born primarily out of their love of and talent for music. Of all his four daughters, three of whom were already married with families of their own, Emily, the youngest, was the most musically talented. Her voice was heaven sent. Everybody said so, especially her father.

Besides teaching piano and organ and giving music lessons out of his home, Edward Cobb served as a musical director at several Portland churches, including the State Street Congregational Church. On occasion he would be called upon to perform at the First Parish Church on Congress Street, the same church where his daughter was scheduled to make her solo debut. People came from all over the state just to hear Edward Cobb play the huge organ at the First Parish Church. His annual Christmas recital at the church was a highlight of the Portland holiday season. There was standing room only when he gave a concert. Edward Cobb spent endless hours tutoring his daughter and even as he grew more ill, slowly wasting away in the throes of tuberculosis, he tried to work with Emily. Toward the end of his life, when he was consumed with coughing fits that barely allowed him to breathe, he would write out his instructions to her. Dutifully, Emily would stand before him in their parlor and practice. He would sit crumpled in a chair, covered with a shawl, sweating and coughing, while Emily practiced. It broke her heart to see him that way, but she couldn't deny him the one thing that seemed to give him joy at the end of his short life; he was only thirty-eight years old when he died of consumption. After his death, Emily didn't sing in public for a year.

From 1851 until 1897, the First Parish Church in Portland employed a full-time musical director, German-born Hermann Kotzschman. On those Sunday's when Kotzschman had other engagements, Edward Cobb was the first person called to fill in. In early 1897, because of financial difficulties at the church, Kotzschman resigned. The musical program was then assumed by Associate Pastor John Carroll Perkins. It was Perkins who convinced Emily to sing again. She began singing in the church choir. It didn't take long for Perkins to recognize Emily's immense talent as a soprano, and he chose her to sing two hymns as a soloist at the Thanksgiving weekend service. Perkins cajoled her into doing it, telling her that above all things, God had given her a talent–a beautiful voice–and that she must not waste it. Perkins told her that he would dedicate the service to her late father if she agreed to sing. Using his considerable influence with well-to-do members of his congregation, Perkins was able to obtain enough money for a round-trip ticket on board the steamship to Boston for Cobb and to pay for her singing lessons. Everything, Perkins told her, had been provided for by the good Lord in his wisdom. Her future, he said, was in the hands of Providence.

Cobb had paid the $2 for a round-trip ticket. She did not have $5 for a stateroom, but it didn't matter. She had reached India Wharf in plenty of time and was ready to board.

19: George Kenniston Jr.

George Kenniston Jr. was returning home to Boothbay after spending Thanksgiving with his older sister in Boston. He was a freshman at Bowdoin College, an all-male college in Brunswick, and had to return for classes on Monday.

He had chosen to go to Bowdoin, rather than Harvard University, which his father, brothers, and uncles had attended, because of Nathaniel Hawthorne. Hawthorne had attended Bowdoin from 1821 to 1825. He was, among other Concord, Massachusetts, authors like Ralph Waldo Emerson and Henry David Thoreau, his favorite author, and George Jr. wanted to emulate him. He did not want to go into banking like his father, George Kenniston Sr., who was the president of the Boothbay Harbor Bank; or politics, like his uncle Albert, who was a state senator; or his two older brothers, who ran manufacturing businesses. He wanted to become a writer.

A shy, retiring young man with delicate features and strawberry-blonde hair like his mother's, George Jr. was scholarly by nature. He was the youngest child, born into a wealthy, well-respected family, and his parents doted on him. His mother had given him a handmade, leatherbound, blank notebook to bring to Bowdoin with him. She had the words "My Life and Times" inscribed in brass on the cover. George Jr. had begun filling up the notebook with observations he had gleaned from the world around him, in the hopes that he would someday rely on the notebook to write his first novel, just as Hawthorne had done while at Bowdoin. Hawthorne had worked on his first novel, *Fanshawe*, while attending Bowdoin and later published it anonymously at his own expense.

Like Emily Cobb, George Jr. arrived at the boarding gate of the *Portland* early and was standing in line. His suitcase had already been taken on board. Dressed in a plaid newsboy cap that his sister had given him as an early Christmas present, a thick wool scarf knotted around his neck, a long tweed topcoat and scuffed Brogans, he carried only his leather-bound notebook with him. In his hand he held his $5 ticket, which would allow him to make the long trip back home in one of the luxurious staterooms on board the *Portland.*

20: Building the *Portland*

There were 163 staterooms on board the palatial *Portland* and each was ornate and pristine. Painted white, with cherry woodwork, they had carved mahogany furniture with wine-colored upholstery and plush carpets. It was by anyone's expectations the lap of luxury. The ship was, as reported in a *Portland Evening Express* article that had been published at its launching, "the finest vessel that will travel the eastern waters."

The *Portland* was built in 1889 by the New England Shipbuilding Company in the seaport of Bath, Maine. The Portland Steamship Company consistently added new steamships to their line to stay ahead of the railroad or other steamship companies, and by 1889 they were in need of a new steamship. At the time, the company was operating three steamships, the *Tremont*, *John Brooks*, and *Forest City*. The line's side-wheeler *Tremont* was very well furnished, but its other vessel, the *Forest City*, was out of date. The *Forest City* had been built in 1854 and its tired accommodations and weak engine meant that it could not satisfy the public's increasing demand for luxury. The company needed a new palatial night boat to maintain its service in the face of greater competition between the Boston and Maine Railroad.

In May 1889 the company contracted with the New England Shipbuilding Company of Bath, Maine, to build a wooden-hulled paddle-wheel steamship to be named *Portland*. In a time when most coastal lines were contracting steel-hulled propeller-driven steamships, the Portland Steam Packet Company chose the Maine shipyards to build the *Portland* because of the low cost and abundance of timber in the area. The *Portland*'s overall design was not chosen for its speed and seaworthiness, but for its shallow draft and the ability to accommodate a great number of passengers in fine style. Its construction cost was a cheap $240,000, a bargain for the time. The *Portland*'s passenger accommodations were what the vessel was known for.

On October 14, 1889, the *Portland* slid down into the Kennebec River to great celebration and cheer. It measured 291 feet in overall length, with forty-two-foot beam, fifteen-foot depth of hold, and a gross tonnage of 2,284. After its launch, mechanics installed the *Portland*'s massive walking beam engine, which had a cylinder diameter of sixty-two inches and a stroke of twelve feet. The single cylinder transferred power via a twenty-foot-long walking beam that in turn, rotated the thirty-five-foot diameter paddle wheels. The steamship reached its top speed of fifteen miles per hour with fifty pounds of steam pressure from its two iron boilers, built by the Bath Iron Works.

The main saloon on the ship was centered on the top deck. The saloon, with its long mahogany bar, mirrored wall, and assortment of comfortable booths and tables, was lit by an overhead skylight that let the sun shine in during the day trips back and forth between Portland and Boston and the moon and starlight on the evening runs. On either side of the saloon were two staircases running to the lower deck lined with glistening and polished brass railings. All of the woodwork inside the ship was intricately ornate.

After the installation of its engine and boilers in Maine, the steamship made its maiden voyage to Boston on June 14, 1890. It returned to Portland that evening, marking the start of its daily trips between the two cities. This first trip was the beginning of the *Portland's* regular schedule, which the steamer maintained until 1898 without much notoriety. Initially, the steamship operated opposite the *Tremont* and then later the *Bay State.*

Captain Blanchard's regular route from Boston to Portland took him out of Boston Harbor, up along the Massachusetts coastline past Salem and Newburyport, and then up past Thacher's Island. From there it was straight on to Portland.

> Winter or Summer. Its steamers are not fair weather or summer craft, but are designed, built, fitted and furnished for all days and seasons, and for every exigency of weather and condition of navigation as well. Summer or winter, spring or autumn–in total darkness or full moonlight–they are the same comfortable, safe, well-ordered and well performing transport agents. Indeed, when warmed and lighted on winter nights, and animated by the presence in their saloons of goodly companies–as satisfied and contented as good cheer can make them–it may well be doubted if the attractions they present in winter are not in many respects superior to those which they hold out in summer time. (George Fox Bacon, *Portland, Maine, Its Representative Businessmen and Its Points of Interest*, 1891)

The *Portland*'s regular voyage from Boston to Maine was memorialized in the poem, "Peggy Bligh's Voyage," by New England poet Lucy Larcom in 1884. Born in Beverly, Massachusetts, in 1824,

Larcom was one of the first teachers (1854) at the Wheaton Female Seminary, now Wheaton College, located in Norton, Massachusetts. The poem memorialized the many ports of call along the coast and was a lasting tribute to the trip's many beautiful sights.

> You may ride in an hour or two, if you will,
> From Halibut Point to Beacon Hill,
> With the sea beside you all the way,
> Through pleasant places that skirt the Bay;
> By Gloucester Harbor and Beverly Beach,
> Salem's old steeple, Nahant's lone reach.
> Blue-bordered Swampscott. and Chelsea's wide
> Marshes laid bare to the drenching tide.
> With a glimpse of Saugus spire in the west,
> And Maiden hills in their dreary rest.

21: Demise of the Clipper Ship

As one thread of this nautical story grew taut, another began to unravel. Although the advent of the steamship for trade and travel was not the *only* reason for the demise of the once dominant clipper ship, it was surely a contributing factor. The boom years of the clipper ship began in 1843, ending in approximately 1870. Although they were much faster than early steamships, they depended on the vagaries of Nature–the wind, tides, and currents–while steamers were propelled by the manmade intricacies of the steam engine, not held accountable to the whims of the elements. Thus, they were better able to keep to a timely schedule.

These tall and majestic ships, with hundreds of yards of sail canvas were built for speed, and the speed of the ships was able to enhance merchants' earnings. They were able to get up to 14 knots or more in a good wind. The very term "clip" came from the phrase, "Going at a good clip," meaning speed, and it was applied to race horses as well as clipper ships.

Clipper ships were used all over the world on trade and passenger travel routes between England and its many colonies with tea cargoes. They also carried trans-Atlantic trade and provided transportation from New York to San Francisco, around Cape Horn, for hopeful prospectors during the 1848 California Gold Rush.

The first American-built clipper ship, the *Ann McKim*, was built in Baltimore in 1833. In the mid-1800s, shipbuilders in Medford, Massachusetts, and soon enough shipyards all along the New England coast began construction of them. Donald McKay is credited with building the fastest clipper ship, the *Flying Cloud*, in 1851 at the Border Street shipyard in Boston. Within six weeks of its launching, the *Flying Cloud* sailed from New York, rounded Cape Horn and made it to San Francisco in eighty-nine days, twenty-one hours, under the command of Captain Josiah Perkins Cressey and his wife, Eleanor, who is credited with serving as the ship's navigator. During those early days of the Gold Rush, it usually took two hundred days to travel the more than sixty thousand miles around the tip of South America. Duncan McLean wrote in the *Boston Daily Atlas*, "If great length, sharpness of ends, with proportionate breadth and depth, conduce to speed, the *Flying Cloud* must be uncommonly swift, for in all these she is great."

Despite this and other great sailing accomplishments of speed, the clipper ship could not pay satisfactory dividends to its owners and merchants based on their construction and maintenance costs to show a long-term source of profit. The factors that caused the end of the clipper ship era were in fact the very same factors that brought their dominance into existence: world-wide trade conditions. Although the clipper ship's speed was of vital importance in trade, transport, and commerce, their cargo capacity was limited and they were, once again, at the whims of Nature. Steamships were more economical and had a larger cargo capacity. Although slow at first, they were soon able to sail at a good "clip." Added to this was the completion of the Suez Canal in 1869, connecting the Mediterranean Sea to the Red Sea. It gave ships a shorter route between the Atlantic and Indian Oceans. This waterway was more easily navigated by steamships than clipper ships. So came the end of the clipper ship's reign over the high seas and the advent of steamships as the principle means of trade, transportation, and commerce.

> The American Clipper has permanent significance for two reasons. First, as already stated, it was the type of vessel by means of which our country rose to her brief period of supremacy upon the high seas. Second, it is one of the most, if not quite the most characteristic expression of our peculiar aesthetic sense which we have as yet produced. It accurately embodies our ideas of beauty, which insist upon thorough utility, the most perfect functionality possible, the greatest possible economy of mere mass, bulk, material, without sacrifice of structural strength, and perfect co-ordination to the environment and the purpose in view. (Charles E. Park, "The Development of the Clipper Ship," The American Antiquarian Society, 1929).

22: The Development of the Steam Engine

The dominance of the steamship for trade, transportation and commerce on American rivers, lakes, coastal excursions and across the seas would not have been possible without the creation and development of the steam engine. The prototype of the steam engine, "the atmospheric engine," was first built by Thomas Newcomen (1664–1729) in 1712. An English inventor, he was an ironmonger by trade (someone who manufactured iron goods). Flooding in coal and tin mines in England, the foundation of iron manufacturing was a major hazard in terms of loss of life and production costs.

Newcomen sought to find ways to pump out water from these mines, which led to his invention of the atmospheric engine, simply referred to as the Newcomen engine. It operated by condensing steam, sucking it into a cylinder, creating a partial vacuum, that allowed the atmospheric pressure to push the piston into the cylinder and drive the workload of pumps and other related machinery. It was the first practical use of steam to do mechanical work. Primarily, these engines were used during the eighteenth century throughout England and Europe to pump water out of mines. In 1765, James Watt (1736–1819) improved the efficiency of the Newcomen engine, doubling the engine's overall fuel efficiency.

Watt was a Scottish inventor, chemist, and mechanical engineer. Working at the University of Glasgow, he became intrigued by the steam engine. He realized the Newcomen engine wasted an abundant amount of energy by repeatedly cooling and then heating the steam-powered cylinder. He added a separate condenser, which avoided the waste of energy. This in turn increased the power, efficiency, and cost-effectiveness of steam engines. Watt ultimately modified his engine to produce rotary motion that allowed its use well beyond merely pumping water out of mines. He entered into a partnership with Mathew Boulton in 1775 to successfully commercialize his steam engine. The company became exceptionally profitable and made Watts wealthy. He went on to develop the concept of horsepower, the unit of measurement of power identifying the rate at which work is done. Watt adopted the term to compare the power output of steam engines with the power of draft horses. The International System of Units (SI) unit of power called the watt was named after him.

23: Development of the American Steamship: John Fitch

The development and success of the steamship belongs to two men—John Fitch and "The Father of Steam Navigation," Robert Fulton. John Fitch (1743–1798) was the first inventor to build a steamboat in the United States. Between 1785 and 1796, he constructed four different steamboats that successfully worked on rivers and lakes.

Fitch's steamship designs used numerous combinations of propulsive styles, including ranked paddles that were modeled after Indian war canoes, paddle wheels, and screw propellers. He introduced the first successful steamboat, *Perseverance*, on August 22, 1787. The forty-five-foot steamship sailed on the Delaware River in front of members of the Constitutional Convention. It was estimated that the ship traveled between thirteen hundred and three thousand miles that summer. The estimated speed of the vessel was a minimum of six miles per hour to a maximum of seven or eight miles per hour, depending on the conditions of the river and the weather.

Fitch went on to build a larger steamboat that carried passengers and freight between Philadelphia and Burlington, New Jersey. He was ultimately awarded his first United States patent for a steamboat in August 1791. He was granted his patent after a legal battle with James Rumsey, who claimed he had built the same steam engine. Fitch built four different steamboats between 1785 and 1796. Each ship demonstrated that using steam for transportation was possible and successful. His ships used a variety of designs, including paddle wheels and screw propellers. Despite his patent and although his steamships proved mechanically successful, Fitch did not pay enough attention to the construction cost of the ships or the overall operating costs, which led to Fitch being unable to adequately justify the financial benefits of steam navigation and ultimately attract investors.

Faced with continued failure to fully realize the successful operation of his steamboat enterprise, Fitch moved to Kentucky in 1797, where he again attempted to work on the development of his steamboat. This too proved disappointing. He began drinking heavily and ended his own life in Juy1798 by taking an overdose of opium. Fitch's idea would be made profitable two decades later by Robert Fulton.

24: Development of the American Steamship: Robert Fulton

"Steam has been applied in America to the purpose of inland navigation with the greatest success."

–*Gentleman's Magazine*, December, 1809

The first successful steamboat was the *North River Steamboat*, which became known simply as, the *Clermont*. It was built by the wealthy American investor and New York politician, Robert Livingston and inventor Robert Fulton in 1807. Fulton (1765–1815) became known as "The Father of Steam Navigation" and is credited with turning the steamboat into a commercial success. Fulton was married to Livingston's niece.

On August 17, 1807, Fulton's *Clermont*, with a complement of invited guests, went from New York City to Albany, making history with a 150-mile trip, taking thirty-two hours at an average speed of about five miles per hour. The ship made one stop at Livingston's estate, Clermont Manor, in southwestern Columbia County. The return trip to New York City was completed in just thirty hours. According to Fulton, "My first steamboat on the Hudson's River was 150 feet long, 13 feet wide, drawing 2 ft. of water, bow and stern 60 degrees: she displaced 36.40 [*sic*] cubic feet, equal 100 tons of water; her bow presented 26 ft. to the water, plus and minus the resistance of 1 ft. running 4 miles an hour."

> Mr. Fulton's ingenious steamboat, invented with a view to the navigation of the Mississippi, from New Orleans upward, sails today from the North River, near State's Prison, to Albany. The velocity of the steamboat is calculated at four miles an hour. It is said it will make a progress of two against the current of the Mississippi, and if so it will certainly be a very valuable acquisition to the commerce of Western States. (*American Citizen*, August 17, 1807)

In 1811, the *New Orleans* was built in Pittsburgh, by Fulton and Livingston. Livingston was a politician and inventor. The *New Orleans* had a passenger and freight route on the lower Mississippi River. A steam ferry-boat was built to ply between New York and Jersey City in 1812, and the next year, two others, to connect with Brooklyn. The Jersey ferry crossed in fifteen minutes, the distance being a mile and a half. Fulton's boat carried, in one load, eight carriages, thirty horses, and three hundred passengers, proving the viability of cargo transport on board a steamship. By 1814, Fulton, together with Edward Livingston (the brother of Robert Livingston), were offering regular steamboat passenger and freight service between New Orleans, Louisiana, and Natchez, Mississippi. Their boats traveled at the rates of eight miles per hour downstream and three miles per hour upstream.

At the time of the War of 1812, Fulton designed a steamship, the *Demologos*, that was capable of carrying a heavy battery of weapons and going four miles an hour. The estimated cost of the ship was $320,000. The construction of the vessel was authorized by Congress in March, 1814. It was the first steam-powered vessel built for the United States Navy. Referred to as *Fulton the First*, it was an enormous boat, with a double hull, measuring 156 feet long, 56 feet wide, and 20 feet deep, and weighing 2,475 tons. Fulton died of tuberculosis at age forty-nine in February 1815 before the ship was delivered to the Navy. After running trials under steam power, the ship was delivered to the Navy in June 1816. Because the war ended in 1815, the *Demologos* never saw action, and no other ship like her was built. It had only a single day of service a year later, when it carried President James Monroe on a cruise in New York Harbor. The ship was destroyed by a gunpowder explosion in June 1829.

> In the *Clermont*, Fulton used several of the now characteristic features of the American river steamboat and subsequently introduced others. His most important and creditable work aside from that of the introduction of the steamboat into everyday use was the experimental determination of the magnitude and the laws of ship resistance and the systematic proportioning of vessel and machinery to the work to be done by them. (Robert Henry Thurston, *A History of the Growth of the Steam Engine*, 1883)

25: Impact on Commerce

Watt's improvements to the steam engine and Fitch and Fulton's development of the successful steamship, ushered in the era of steam to power boats previously driven by wind alone. Steam power revolutionized transportation on both sea and land. Its development drove the Industrial Revolution, which led to the use of powerful locomotives that crisscrossed America and created cargo and passenger ships used along the country's many rivers, lakes, and coasts as well as in trans-Atlantic and -Pacific voyages.

These early steam-powered engines, were rudimentary, inept, massive, and inefficient. The building and installation of increasingly more efficient engines for ships took place in England, France, and the United States at a rapid pace during the mid-1800s.

> Unlike muscle power, it never tired or slept or refused to obey. Unlike waterpower, its immediate predecessor, it ran in all seasons and weathers, always the same. Unlike wind, it responded tractably to human will and imagination: turning on and off, modulating smoothly from the finest delicacy to the greatest force, ever under responsive control. It is impossible to contemplate, without feeling exultation, this wonder of the modern art. (*The Quarterly Review of London*, 1830)

The growth of steamship service in America was a boon to commerce throughout the country. Long before current modes of transportation including the automobile, trucks, trains, and airplanes, it was the navigation of America's rivers that connected the country and were used to move freight and passengers. Travel on rivers was slow and tedious and had for many years depended exclusively on the current of the rivers and the use of human strength. All of that changed dramatically with the advent of the steamship in the early 1800s. According to Robert Henry Thurston, author of *History of the Growth of the Steam Engine* (1883), "We must return to America to witness the first and most complete success commercially in the introduction of the steamboat."

Steamboats were able to travel America's rivers at the incredible speed of nearly five miles per hour and were no longer dependent on a river's current or on manpower. For decades these boats dominated the process of river travel for both passengers and goods. There were, of course, risks associated with steamboat travel, including explosions on board and dangerous weather conditions. In April 1838, the 150-ton steamboat *Moselle*, heading east on the Ohio River, exploded, spewing steam, boiler parts, and fragments of bodies. Within fifteen minutes of the explosion, only the smokestacks and a segment of the upper decks still showed above the water. Rescuers could only save about half of the passengers, and many who were not killed by the initial blast drowned in midstream. All told, about half of the 280 people on the *Moselle* died. It was the worst steamboat disaster ever at that time. Eventually, other forms of transportation were introduced and became more reliable and faster, but for decades the steamboats reigned supreme as the nation's system of travel.

26: River Steamboats

Steam-powered boats would soon dominate the waterways and forever transform river and coastal travel and trade. Steamships played a vital role in the expansion of the United States. The years after the Revolutionary War westward expansion was facilitated by routes along southern rivers like the Mississippi, Apalachicola, Chattahoochee, Alabama, and others. In 1803, through the Louisiana Purchase, America acquired the city of New Orleans and the large swaths of Louisiana territory. The rivers flowing through the region allowed settlers to move west. An abundant number of cities sprang up along the rivers to make transportation and trade easier. Only flat-bottomed keelboats were carrying goods and passengers along these plentiful rivers and their speed depended on the river current. Downstream was easy. Heading back up stream rivermen had to pole the boats against the current. A round trip could take as long as nine months. Following Fulton's successful steamboat excursion in 1807, steamboats began to take over the role of the primary source of river transportation.

Most steamboats were paddle wheel boats. They had a basic design including a hull, or body, made of timber and later steel, and a wooden paddle wheel. The paddlewheel had a spherical center with shafts coming out of it much like a bicycle wheel. Planks were attached to the spokes to make the paddles. These huge paddles were placed on either the side or rear of the boat. Boats with paddles on the side were called side-wheelers. Boats with a paddle at the rear were called stern-wheelers. The paddle wheels were run by an engine that was powered by steam, which was generated by coal- or wood-fired boilers.

Various steamboats had different jobs. Towboats moved barges by pushing them up and down river. Ferries carried people across rivers. Snag boats cleared the river of debris dangers. Packets carried provisions and people. The most famous type of steamboat along the rivers and whose style was later adapted for coastal and ocean travel, was the showboat. Showboats were the floating palaces of the nineteenth and early twentieth centuries. Many showboats were beautifully decorated and had theaters, galleries, ballrooms, and saloons.

According to *A History of Steamboats*, by the United States Army Corps of Engineers, these showboats, "traveled up and down rivers bringing plays and music to river towns. Showboats would announce their arrival by playing their organ-like steam calliope, which could be heard for miles. While showboats provided excitement and entertainment for river towns, they were never very common. In 1900, there were less than 30 showboats, and by 1930 there were less than 10."

Two of the biggest dangers to steamboat passage along inland waterways were from Indian attacks and boiler explosions. Indians would hide along the banks of a river and attack. If the ship wrecked near the bank, it would lose its cargo, "and the crew and passengers might even lose their lives. Indian attacks were a concern, but the biggest danger facing steamboats was boiler explosion." If boilers were not carefully watched and maintained, pressure could build up in the boiler and cause a spectacular and deadly explosion. According to reports, from 1811 to 1851, 21 percent of river accidents were caused by explosion.

> Steamboats were used in United States so widely, that by end of mid 1850s there were as many as 1200 boats sailing.
>
> –Louis Hunter, *Steamboat on the Western Rivers*, 1949

27: Trans-Atlantic and Coastal Passage

The use of steamships like the *Portland* evolved to coastal and trans-Atlantic passage. In 1819, the *Savannah* made the first Atlantic crossing powered partially by steam. In 1838, the British and American Steam Navigation Co.'s *Sirius* left Ireland with forty paying passengers for a historic voyage to New York. It took eighteen days and the *Sirius* ran out of coal–the crew had to burn the cabin furniture and even a mast–but it was the first passenger ship to cross the Atlantic entirely on steam power. The *Great Western* steamship left Bristol, England, four days after the *Sirius* and arrived in New York Harbor only four hours behind it, making the crossing in fourteen and a half days. There was a first-class deck where travelers voyaged in comparativc luxury, while those who could not afford first class, made the journey in cramped quarters on the lower decks. Former reliance on wind and weather for the sailing ships was superseded by dependence on fuel sources–first the burning of wood, then coal, and finally oil. Steam power promised to free ocean vessels from the whims of wind and weather. But steamships suffered from a number of problems, including carrying enough fuel, finding reliable engines, and supporting huge operating costs. Early steam vessels often relied on both steam engines and sails. By the 1850s, many wealthier passengers moved to steamships for transportation purposes, while most immigrants still crossed the ocean on sailing vessels.

Shipping along America's coasts was vital to the nation's economy, and steamships began plying their trade all along the New England coast to accommodate the need for supplies and material. Lumber, bricks, cotton, and other bulk cargoes from different parts of the country spent time at sea. Many American cities were built with materials carried over coastal waters. Limestone quarried in Maine was made into mortar and shipped to New York and Boston, where it was used in building construction.

Passenger travel between major ports like the Boston-to-Portland line by the *Portland* quickly came into vogue. Other noted coastal steamship excursions included the Fall River Line established in 1847, which provided steamship transportation from Fall River, Massachusetts, through Narragansett Bay and Long Island Sound into Manhattan.

28: Indecision

> The fishermen know that the sea is dangerous and the storm terrible, but they have never found these dangers sufficient reason for remaining ashore.
>
> –Vincent Van Gogh, 1853–1890

Hope Thomas stood in line waiting to board the *Portland* holding on to her two bags filled with Christmas presents. She was bundled up in her long woolen coat and scarf with a gray, silk, spoon bonnet, a black velvet ribbon securing it and tied in a bow beneath her chin. The electric lights along India Wharf had been turned on and brilliantly lit the entire length of the huge pier. The lights on board the *Portland* had also been turned on, illuminating the luxurious ship in celestial brilliance.

Built in 1807, India Wharf was one of the largest commercial wharves in the Boston port. It included a handsome five-story brick façade building that ran the length of the pier with a slate roof, ornate stone trimming on its many windows, marble appurtenances, and red sandstone adornments. The foundation was made of stone-gray granite and the floors were made of fine hewed pine. It boasted more than thirty stores for shopping. It was a hubbub of noisy activity that night with men, women, and children, coming and going—some, like Hope, getting ready to board, others there to see friends and family off. There were ship crews making their way along the cobblestone paved surface that lined the pier and stevedores busy loading vessels for departure.

Hope was the oldest of the three Winslow sisters. While she had married Bill Thomas, a fisherman from Maine, her younger sister Grace had married a lawyer from Boston and lived there now in a brownstone along Beacon Hill with her husband. They had no children. Her youngest sister, Charity, was not married and lived with Hope in Portland, where she cared for their sick mother and took care of Hope's two young children when Hope was off to visit her sister in Boston and Bill was away at sea. The Winslow sisters remained close. Hope telephoned her sister in Boston at least once a month, wrote letters and cards and came to visit whenever she could. Coming down on the Thanksgiving weekend to visit and shop with her sister was a yearly ritual.

In the bags she carried with her, Hope had a new Cavendish pipe she had purchased at the Boston Smoke Shoppe along Charles Street. Bill Thomas had chewed the stem of his old pipe down to nothing. She had bought him a pouch of tobacco as well. For her oldest daughter, ten-year-old Victoria, she bought a copy of Andrew Lang's best-selling book, *The Pink Fairy Book*, which had just been published last year. She had bought a copy of Lang's first illustrated book of fairy tales, *The Blue Fairy Book*, the year Victoria was born, 1889. It contained one of Victoria's favorite stories, "Rumpelstiltskin." It was the first time the classic story appeared in English and it was beautifully illustrated. Each year that Lang published another of his fairy tale collections, Hope bought a copy for her daughter. So far, Hope had bought the *Blue* (1889), the *Red* (1890), the *Green* (1892) and the *Yellow* (1894) fairy books by Lang. It was the perfect gift for her studious daughter, who savored all of Lang's tales, even more so since she had learned how to read. But even before then, Hope would read the stories aloud to her. For her youngest daughter, Eliza, five years old, she bought a stuffed bear. Eliza liked to cuddle with stuffed toys, so much so she had nearly squeezed the stuffing out of her stuffed toy rabbit.

For Mother Winslow, Hope's eighty-year-old invalid mother, in whose old Federalist farmhouse in Bailey's Island she, Bill, her two daughters, and her sister Charity all lived she bought a tin of Oolong tea. For her younger sister Grace, she bought a linen tea wrapper dress made of an ecru cream-colored natural linen fabric, with a white floral embroidered scalloped trim edging and a white eyelet lace collar and cuffs. It was a beautiful gift, and she knew Grace would love it. She could wear it when entertaining many of John's clients and guests. It was a fabulous floor-length one-piece wrapper style dress with a full skirt, long train, and bustle back, with straps sewn inside for fullness to accommodate the bustle. The dress had large mother-of-pearl buttons down the front. Hope wished she could see her sister's face when she unwrapped the gift at Christmas, but that would be impossible. She would be back home in Portland, and Grace would be there in Boston on Christmas day. Hope had wrapped the dress and given it to Grace before she left. For her youngest sister, back home in Portland, she bought a fur muffler and matching hat. And for her brother-in-law, John Pendleton, she bought a copy of the best-selling book, *Quo Vadis*–the book store had a colossal book display featuring Henry Sienkiewicz's 1896 historical novel selling at half the price.

She had come to Boston on Wednesday to help her sister prepare Thanksgiving dinner, while her husband Bill had put out to sea. Charity would prepare Thanksgiving dinner for her mother and the children back home in Maine. Hope ate dinner with Grace and John and his parents on Thursday and on Friday the two sisters went shopping.

Along with the presents for everyone else, Grace, at her sister's urging, bought a pair of new shoes for herself. She couldn't make up her mind whether to buy the black leather high-top button boots or the two-tone soft leather shoes. They both fit her so well. One, the black leather high-tops, were utilitarian. She would get a lot of use out of them back home where she took care of house, cooking and cleaning and caring for her daughters, her fisherman husband, and her ailing mother. The black boots had square toes and a leather front with a black wool upper boot. There were nine black buttons along the scalloped sides and wooden round cube heels.

The other pair, the two-tone shoes, were more fashionable, made for a summer promenade or better yet, to wear to one of the church socials. They were made for dancing, she was told by a clerk. They had ten buttons, a pointed toe, and were made of soft brown and tan leather, with low stacked heels. Both pairs flattered her slim feet. She knew which pair her sister wanted her to buy. She could hear her sister's voice now: "Hope, buy the two-tone. The others are so fuddy-duddy. You're not a hundred years old, don't you know? Have some fun."

Fun, Hope thought, had not been in the equation that was her life in Maine for many years; not that her husband, Bill, hadn't been a good husband and provider; not that she didn't love him with all her heart and he her; not that she didn't adore her daughters; and not that she didn't freely care for her ailing mother, but things like the annual summer promenades and the spring and harvest dances at her church were things of the past for her. There was just too much work. Still?

She knew what she should do, keep the black shoes, but there was still a nagging doubt.

Her sister and her husband had ridden down to India Wharf in a carriage to see her off. As the horse-drawn carriage neared the wharf she could see a crowd forming at the single ticket office along the pier where the *Portland* was docked. She heard the steamship whistle blow as the carriage came to a stop alongside the gangplank where passengers were getting ready to board. The driver helped her out and handed her the two shopping bags and umbrella. After a round of hugs and kisses and relentless goodbyes during which Grace burst into tears at saying goodbye to her sister once more, they left Hope standing in line at the pier waiting to board.

She looked up the lighted busy thoroughfare and thought about the new shoes she had purchased. The store where she bought them was quite a trek away. If she went back, she might miss boarding and would have to stay until Sunday. She could hear the roar of the ship's huge engines. Foam began to bubble up from the ship's propellers.

The two side paddles began to creak. There was a loud blast from the huge steam whistle atop the *Portland*. It was time to board.

29: Telephone Contact

At around 5:30 p.m., John Liscomb, the new general manager of the Portland Steamship Company, telephoned the Boston steamship office from Portland and spoke with the company's Boston agent, Charles Williams. He asked to speak with Captain Blanchard, but Blanchard was not at the office at that time. Liscomb reportedly told Williams that, based on weather forecasts, Blanchard should wait until 9 p.m. before leaving Boston, and that if the weather continued to worsen, Blanchard should cancel the trip back to Maine. Liscomb informed Williams that the *Bay State* would not be making its regularly scheduled run from Portland to Boston because of the ominous weather forecast. According to Williams, Liscomb's message was given to Blanchard when he returned to the steamship offices later that day.

"I don't understand what they are afraid of," Blanchard said. "I can take this ship through any storm without fear." Despite the instructions from Liscomb, Blanchard informed Williams that he intended to leave Boston Harbor on time that night.

Around 6 p.m. Blanchard received a telegram containing the weather report from New York. The telegram stated that it was snowing in New York but that the wind direction had shifted to the northwest. This telegram may have led Captain Blanchard to believe that the most severe portion of the storm had passed and that the storm was moving out to sea. Since he was headed northeast, Blanchard would be headed away from the storm, unlike the *Bay State*, which would have been heading into the storm on her trip from Portland to Boston. It was nearly 7 p.m., and although the sky was overcast, there was no snow and only mild wind.

30: Crash Course

Weather reports coming out of Washington and Boston changed as rapidly as the weather itself. According to various weather reports gathered in the Washington, D.C., Weather Bureau's home office, a high-pressure system had settled across Ohio and the Midwest, making that entire section of the country enveloped in a relatively seasonal warm front. Chillier winds were blowing in from the Great Lakes region followed by what most observers considered trifling snow flurries and brisk gusts of wind and barely noticeable accompanying showers. There was no cause for alarm at the Weather Bureau's main offices.

Captain Blanchard had studied all the available weather forecasts that day and knew of the predicted storm front closing in from the Great Lakes region. He had met with John W. Smith, the chief meteorologist at the Boston Weather Bureau office earlier in the day to go over the most recent weather forecasts. He learned from Smith that the storm bearing down from the Great Lakes region was making its way toward the New England coast on a path that would intersect with a second storm that was moving up the coast from the south. Based on this information, a special weather advisory was issued by the Weather Bureau's Washington, D.C., headquarters calling for heavy snow fall and gale winds all along the New England coast beginning on Saturday evening and lasting into late Sunday.

The initial Weather Bureau's forecast for Saturday and Sunday gave Blanchard no real reason for alarm. The forecast issued on Friday, November 25, called for, "fair continued cold, brisk westerly, shifting to southerly winds," for New England. For New York the forecast called for, "increasing cloudiness and rain or snow by Saturday night." According to the forecast, "The weather in New England Sunday will be unsettled, and rain or snow is possible, most likely in the afternoon or night. . . . The temperature will not change decidedly."

The Saturday weather forecast didn't change dramatically, until later in the day when a special weather advisory was issued alerting maritime and other transportation concerns of higher than previously reported wind velocity and snow fall. According to the weather advisory issued late on Saturday, November 26, Maine, New Hampshire, and Vermont could expect "heavy snow" on Saturday and "snow and much colder; southeasterly winds shifting by tonight to northeasterly gales." For Massachusetts, Rhode Island, and Connecticut heavy snow was predicted with clearing on Sunday followed by much colder weather. According to the advisory, there would be, "northeasterly gales tonight, and northwesterly gales by Sunday."

31: Rollinson

Eben Heuston went about his duties as chief steward as he usually did before the ship's departure, making sure the staterooms were ready, deck chairs lined up, and blankets provided. Everything about him was immaculate. His curly, nearly all white hair was always brushed back in a perfect sheen, parted expertly down the middle of his head–the part so straight and level you might think he had measured it with a ruler and level. His moustache was neatly trimmed. His fingernails were cleaned and manicured. His uniform was clean and pressed. He took care of washing and ironing his steward's uniform himself. You could slice bread with the crease in his pants. His cuff links sparkled and his shoes were polished. Some people might have imagined he was so meticulous about his appearance because he worked on the *Portland*, but whatever job he had, Heuston would have been equally immaculate about his appearance. He always had been.

Like most of the crew on board, Heuston was aware of the storm that was being predicted. He had sailed with Captain Blanchard many times before during the past three months and had the utmost faith in the Blanchard's seamanship. If it wasn't safe to leave port, he knew Blanchard wouldn't go. Above all else, he knew Blanchard was a cautious man. As Heuston went about his assigned chores before the passengers came on board, he turned the corner just outside the galley and ran into Harry Rollinson. Rollinson was one of ship's engine stokers, responsible for feeding the ship's huge furnace with coal and stoking the roaring flames. Tall and muscular from years of hauling and shoveling coal in the boiler rooms of a hundred different ships like the *Portland*, he wore the uniform of a working man, denim bib-overalls and denim shirt, shirt-sleeves rolled up high onto his muscular arms. A red bandana was tied around his thick neck. Nothing about Rollinson belied his age of fifty-two. There was not a wrinkle on his face or a bit of fat on his body. He was all sinewy muscle. Unlike his friend Eben, there was not a speck of gray in his hair.

Rollinson wore a gold earring in his right ear. The earring spoke of a wilder period in American seamanship when a pierced earlobe was a symbol that the wearer had sailed around the world or had crossed the equator, which Rollinson had done on several occasions before joining the crew of the *Portland*. And for many seamen like Rollinson, the earring would be used as payment for a proper Christian burial, should they drown at sea and their bodies be recovered. The gold earring would cover burial expenses.

Heuston didn't wear an earring. He told his friend it was because he did not intend to die at sea. Although friends for many years, Heuston and Rollinson were as alike as night and day. Heuston never smoked, drank, or swore, proclivities that Rollinson often engaged in.

"What are you doing here? Captain will be calling for you," Heuston said, knowing Captain Blanchard would be calling down to the boiler room soon looking for his engine room crew to build up a head of steam before they left the harbor.

"Don't matter," Rollinson said striking a match against the wall he was leaning against and lighting a cigarette that dangled from his lips.

"Better put that out in case he sees you," Heuston warned him.

"Can't smoke once I'm down in the hole," Rollinson said. He took a small flask from his back pocket, opened it, and took a swig of the whiskey that was inside it.

"Don't let him catch you," Heuston again warned him.

"He don't care about me now the ship's going. He just wants to strut around in front of the passengers to impress them. You show me again before I get below."

"Last time," Heuston said and reached into his pocket. He pulled out a small velvet covered box and snapped it open. Inside was a solid gold cross on a delicate gold chain.

"That is something beautiful. Your Margaret is gonna be mighty happy with you," Rollinson said smiling broadly."

"One year ago today," Heuston said.

"You're a lucky man," Rollinson said slapping his friend on the back and taking another swig of whiskey before putting the cap back on and placing it back into his pocket.

Rollinson was unmarried. He lived at home where he cared for his ninety-two-year-old mother. And although he was anything but a good Christian living soul, he brought his mother to church faithfully each week.

Rollinson flicked his lit cigarette over the side and slapped Heuston on the back once more.

"You think Blanchard's gonna sail with this storm coming?" he asked Heuston.

"Blanchard's a good captain. He'll do what's best," Heuston said.

"What's best for him," Rollinson said.

"You don't have to worry about the storm," Heuston assured him.

"Well, I ain't going down if it happens. I'll take my chances on the open sea. See you in port," he said.

Heuston went back to his duties as Rollinson disappeared down one of the narrow spiral staircases leading down into the bowels of the boat.

Of captain brave, he was the best,
To my aye storm a-long!
But now he's gone and is at rest;
Aye, aye, aye, Mister Storm a-long.

(sea chanty of African American origin, 1840)

32: All Aboard

Charlie Thomas ran a small grocery store in Deering, Maine. He; his wife, Susan; three-year-old daughter, Gladys; and their small dog were getting ready to board. Thomas, no relation to Hope Thomas, had reserved a stateroom for himself and his family. Thomas heard about the storm that was approaching but put his faith in Captain Blanchard. He knew Blanchard and his wife since they frequently shopped at his store. Thomas knew Blanchard was a seasoned sea captain. If he thought the *Portland* and its passengers would be in any danger, he would cancel the scheduled run. He also knew the young captain of the *Bay State*, Alexander Dennison. Dennison was young and new to his command. Thomas heard that Dennison had canceled his trip from Portland to Boston that night. He figured the young captain was just being cautious. Dennison wasn't as experienced as Blanchard and probably didn't know the waters as well. The threat of the storm had probably spooked him.

Thomas and his family were getting ready to board when he heard the man in line in front of him grumbling about the weather. E. Dudley Freeman was a former state senator from Maine. Freeman said that he had heard that the *Bay State* was staying in port in Maine because of the reported storm warnings. He said he prayed the *Portland* wasn't going to do the same. He wanted to get back home.

"Of course the ship's leaving tonight," Charlie Thomas interrupted Freeman.

"Are you sure?" Freeman asked quizzically.

"Of course," Thomas told him. "You can be sure of it. Captain Blanchard's no chicken."

"Chicken? You're right. He's no chicken," Freeman said and laughed.

"You think Blanchard's afraid of a little storm. It hasn't even started. You think he would risk heading back to Maine if it wasn't safe? Of course, we're going," Thomas assured the former senator.

George Kenniston Jr. stood in line behind Thomas and overheard the entire conversation. He was relieved.

Emily Cobb was anxious to get on board. She stood behind Kenniston shivering. She was wearing only a thin cotton coat and a shawl over her head. The temperature had dropped and snow flurries began to swirl harder than before. The pier was noisy and crowded. She accidently dropped the satchel she was carrying and some of her sheet music fell onto the cobblestones. Kenniston knelt to help her pick up the sheet music, noticing several of the titles.

He handed the sheet music back to her, noting he was a fan of one of the hymns she had dropped, "Softly Now the Light of Day," written by New Jersey Bishop George W. Doane. Cobb thanked him. Kenniston could see how red and chaffed her hands were. She had no gloves.

KENNISTON.
Are you a musician?
COBB.
I'm a soprano at the church.
KENNISTON.
What church?
COBB.
The First Parish Church in Portland. I'm giving my first recital on Sunday. I have to get back to Portland tonight.
KENNISTON.
Don't worry. You'll get back home in time.
COBB.
Do you think so? We're going to sail aren't we?
KENNISTON.
You can be sure of it. Captain Blanchard's no chicken.

"Softly Now the Light of Day"

Softly now the light of day
Fades upon my sight away
Free from care, from labor free
Lord, I would commune with Thee.

(George W. Doane, 1824)

33: All Ashore

The line of passengers continued boarding. The thunderous boarding steam whistle on the *Portland* blew, reverberating the length of India Wharf. Passengers settled comfortably into their cabins while the crew scurried on deck securing lines; stewards and porters served food and drinks; and below, in the engine room, crewmen heaved shovelfuls of coal into the raging boilers of the steamship. A flurry of snowflakes fell onto the ship's finely polished wooden decks. Captain Blanchard stood on the bridge of the steamship issuing orders below for more steam.

The voyagers crowded into the foyer on the top deck and filed into the dining room and saloon on the lower one, where it was warm, and into their staterooms. The main dining room and saloon was a beehive of noise and activity, with men scurrying around, tables set by porters, food served by waiters and waitresses, liquor from the ample bar served across the mahogany bar or carried across the room in crystal glasses atop silver trays. A small buffet table was set up in the center of the room where children picked at slices of ham and turkey and others munched on warm bread and butter, some sipping hot coffee or cocoa or hot rum toddies.

Eben Heuston continued on methodically about his duties, stopping at cabins, helping with baggage and bundles or bringing out blankets for those who chose to stand outside on the main decks or sit in the row of deck chairs that were all arranged neatly in rows along the top and lower decks.

George Gott stood in line waiting to board. Gott was from Brookline, Massachusetts. A shoe salesman by trade, he was short and plump; his button nose was rosy from the cold and his round cheeks looked like red apples waiting to be picked. A gust of breath tumbled out of his mouth as he spoke. Hope Thomas was in line behind him. Gott turned to her, tipping his bowler.

"There's a rumor of one coming up the coast," he said. "I'm thinking of waiting myself. There' something going on and I don't like it."

Gott pointed in the direction of the huge knotted rope along the bow of the steamship where a black cat was systematically removing her litter of kittens. The cat carried them one at a time in her mouth along the rope and set them down in an empty crate along the wharf. Hope became mesmerized by the way the cat slinked along the rope carrying her litter off the ship.

"By God, if that isn't a sign," Gott said. "Any cat that decides to abandon ship is a good enough reason for me."

Gott tipped his hat again at Hope and waved his hand in a gallant gesture in the opposite direction of the shuffling crowd.

"Would you like to leave as well?" he asked.

Hope fumbled for an answer. She stared back to where the black cat had been, but it was gone now along with all her kittens. The ship's steam whistle blew again. It reverberated deep in her stomach. She looked up at the sky. It seemed even the clouds had clouds. There was no moon or stars. It had become increasingly colder and the snow flurries had turned into a full-fledged snowfall.

"I'll be off then," Gott said. He turned on his heels and pushed his way out of the boarding crowd and disappeared up the pier.

An old seafarer, seventy-eight-year-old Lars Olsen, was reported to have missed boarding the *Portland* by seconds because he left the wharf for one last hospitable drink ashore with friends and missed the boat's departure. It was surely a lucky drink for the old skipper.

34: Superstitions

George Gott's decision to cancel his trip home on board the *Portland* was based on the amalgamation of two different seafaring superstitions–rats always deserted a doomed ship and black cats were a bad omen because they were the pets of witches–but Gott had plenty of superstitions to choose from. The seafaring world is overflowing with them.

One of the most common beliefs was that any ship that set sail on a Friday would encounter an unlucky voyage. Ships also rarely put to sea on a Sunday since it was the Lord's Day. Those who were out at sea on Sunday were expected to observe religious ceremonies.

Whistling or singing into the wind was believed to "whistle up a storm," and bring on sudden gales.

Hammering a nail on a Sunday while at sea was believed to bring tragedy and misfortune.

A single bird landing on board a ship was viewed as a bad omen and many ships would head back into port if such an ominous incident occurred. Birds were thought to carry the souls of dead sailors. Killing a dolphin, a gull, or an albatross brought bad luck. The albatross story was immortalized in Coleridge's "Rime of the Ancient Mariner," in which a sailor killed the bird and was dogged by calamity.

If any odd thing happened during the launching of a new ship it was viewed as a bad omen. Many times, crews refused to sail on ships that were rumored to have had bad launchings. If word got out about a particularly bad launching, owners even had a hard time selling the ship.

The most common superstition was that rats always deserted a doomed ship. This was somewhat founded in truth since rats do not like to get their feet wet and will abandon a ship when a leak happens in the hold of a ship.

Horseshoes on board ships were considered good luck as long as they were nailed in a position with their points facing up.

Flowers weren't welcomed aboard ship because of their association with funerals. If flowers were brought on board as a bon voyage gift, they were quickly thrown overboard.

Ringing bells were also thought to be a bad omen. Bells were also associated with funerals and were thought to forecast death. The ringing associated with the clinking of a wine glass was such an unwelcomed sound and had to be stopped before its reverberation ended. Ship's bells, however, were exempted from this superstition, because they signaled time and the changing of watch duties. But if they rang of their own accord, as in a storm, it was believed someone was going to die.

Although all these many superstitions might have played a role in *Portland* passengers like George Gott, changing their plans, it was more likely that the issue of the weather played a far more important role. But still and all, even that didn't seem to have as much an impact on many of the ship's passengers who desired to return home to Portland above all else–to be home with their families and share Thanksgiving stories in front of a warm familiar fire.

35: The Telegram

Anna Young of Boston was already on board the *Portland* with her young daughter headed to Maine to visit with her mother when a ship's steward knocked on her stateroom door and delivered a telegram for her. She hurriedly opened the telegram. It was from her mother in Portland. She read it quickly. Her mother pleaded with her to wait in Boston for another day. She wrote that there were reports that a big winter storm was predicted to hit the coast. She begged her to disembark and wait out the storm in Boston.

Young looked outside her stateroom window. The sky looked threatening. There wasn't a star visible in the sky. She knew how much her mother wanted to see the baby. She wouldn't have done anything to keep from seeing her unless it was important. Young quickly bundled up her baby daughter, grabbed her bag, and bolted out of the small stateroom leaving the cabin door ajar. She headed straight toward the gangplank pushing past the boarding passengers, marching against the crowd, trying to get off the ship. She clutched her baby, bundled up in a blanket, to her chest with one hand, while she dragged her suitcase with the other. She maneuvered awkwardly down the gangplank and off the ship.

"Carrying my child, I ran for the gangplank just as they started to lift it, and they waited for me," Mrs. Young later said. "When I got ashore I heard the final whistle of the *Portland* as she left the wharf."

36: Another Motive

I must go down to the seas again, to the lonely sea and the sky,
And all I ask is a tall ship and a star to steer her by.

(John Masefield, "Sea Fever," 1902)

Another motive that was later attributed to the reason why Blanchard left Boston Harbor in the face of the coming storm was promulgated by Carrie Courtney, a young woman who had made the trip down to Boston on board the *Portland* and was scheduled to make the return trip home that evening. Courtney spoke with Blanchard shortly before the ship left Boston Harbor. According to her recollection, Blanchard proudly showed her an expensive wrist watch he had purchased for his daughter and explained to her his intent to give his daughter the gift at a party his wife had planned.

According to Courtney, it didn't sound like Blanchard had given serious consideration to staying in port to ride out the pending storm.

"He was more than a little put out for he was giving his daughter a coming out party the first of the week and showed me the beautiful watch he had for her," she later said. "The boat was pitching badly even then."

Courtney decided against taking the steamship home and instead she booked train passage back to Maine.

37: Permission to Stay Ashore

First Pilot Lewis Strout; First Mate Edward Deering; and the clerk to the ship's purser, John Hunt, were all given permission to stay ashore in Boston that Saturday to attend the funeral of Captain Charles Deering, who had been a longtime captain with the Portland Steamship Company. Fatefully, they were not aboard when the *Portland* left Boston for Maine.

Lewis Nelson took over as pilot for Strout, while John MacKey assumed the duties of Deering. Purser Frederick Ingraham worked by himself. Strout and the others planned to return to Portland the next day on board the *Bay State*. Strout was far more comfortable with the *Bay State*'s captain, Alexander Dennison. He had attended the Massachusetts Nautical Training School in Boston, unlike Blanchard, who had come up through the ranks. Dennison was less rigid and far more personable than Blanchard. And as far as Strout was concerned, he was a better captain.

38: Time and Tide Wait for No Man

At approximately 7 p.m. on Saturday, November 26, 1898, Captain Blanchard ordered the steamship on its way out of Boston Harbor. As the Portland steamed out of Boston Harbor, Eben Heuston methodically went about the business of helping passengers settle in for the long trip home. The *Portland*'s whistle blew as the ship churned out of the safe harbor headed toward open sea. As the Portland made its way slowly out of Boston Harbor on its return trip back to Maine, a succession of smaller fishing boats sailed past it, heading back into the safety of the harbor, seeking shelter from the impending storm.

Captain Blanchard stood issuing orders below to the engine room, calling for more steam. Below, Harry Rollinson shoveled mounds of coal into the raging furnace. The luxurious steamship *Portland* steamed along through the choppy waves without incident. Nothing seemed remotely out of the usual. Passengers wandered outside, on deck, children played, couples strolled arm-in-arm, some stood along the railings looking out as Boston Harbor began to fade from sight. The lights from the city twinkled and grew distant. The huge smokestack in the middle of the ship blew off a cloud of black smoke billowing up from the roaring engine below. There wasn't a single star visible, and the moon was hidden behind a layer of rolling clouds.

No one will ever know why Captain Blanchard decided to make the return trip that evening or why he didn't turn the *Portland* around and make for safe harbor when it became clear to him that he was sailing straight into a furious winter storm. Portland Steamship Company officials later claimed that Blanchard had been contacted by telephone and explicitly told to remain docked at Indian Wharf in Boston and wait out the storm. It seemed unlikely that had Blanchard received such a message from the company that he would have disobeyed orders. Others contended that Blanchard was never contacted by the steamship company, and if he had been, he was ordered to set for home that evening.

What was clear was that Blanchard checked the weather reports prior to his departure and it may just well have been that he was confident that, regardless of the forecast, he could stay ahead of the storm and make it safely back to Portland Harbor. For whatever reason he steamed out of Boston Harbor that evening, the tragic result was the same.

> What to them we seem to hear him say, those stormy seas and cruel tornados, those sinking ships and praying hands? Downwards from their unexhausted sources flow the streams through time-worn channels to a changeless sea, a sea whose shores are strewn with the wrecks of empires. (George Granville Bradley, *Lectures on Ecclesiastes*, 1898)

BOOK TWO

The Storm

"All agreed that the storm was one of the worst known . . ."

–*Boston Herald*, December 1, 1898

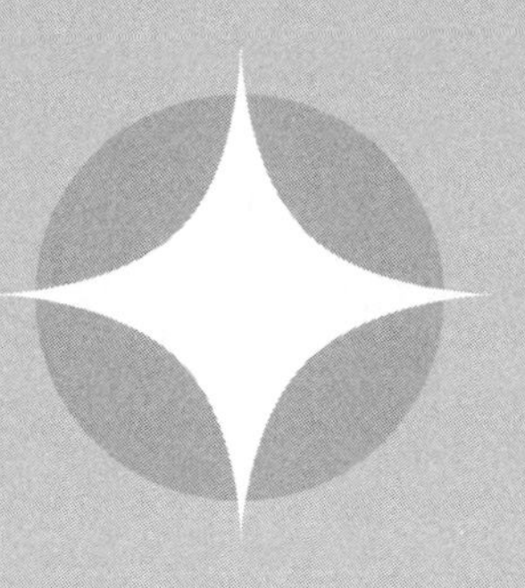

1: The Storm Begins

The storm is up, and all is on the hazard.

–William Shakespeare, *Julius Caesar*

Captain Blanchard's regular route back to Portland took him out of Boston Harbor, up along the Massachusetts coastline past familiar ports like Gloucester, Salem, and Newburyport, and then up past Thacher's Island. It was no different on that Saturday evening. The *Portland* churned along through the choppy waves without incident. Nothing seemed remotely out of the ordinary.

Although neither Boston nor the Maine coast had felt the full effects of the gathering storm, by the time Blanchard steamed out of Boston Harbor, Cape Cod was being rocked by wind gusts approaching nearly ninety miles per hour, a hurricane force. The snow pelted the Cape coastline and inland. Cape Codders knew they were in for a storm of what looked to be epic proportions. They prepared, locking up farm animals, securing store fronts, stocking up supplies and shuttering their homes. Despite their best efforts, Cape Codders had no idea what they were in store for, no one did. Weather forecasts for Saturday and Sunday had given no real reason for alarm and had not predicted the severity of the storm that would become known as, "The Portland Gale,"–the storm of the century.

Almost everyone connected with the steamship company knew Captain Blanchard's inclination toward safety and were comforted in the presumption that, given the treacherous weather conditions, Blanchard surely would head for safety in some familiar port along the way back to Maine if his judgment indicated it was necessary.

> I was on Cape Cod Tuesday after the storm and talked with many of the life-savers and others who were out in the blow, and they all seemed to agree that nothing so severe has ever been experienced in that part of the country. (Charlie Ward, *Boston Herald*, December 3, 1898)

2: Sightings

Other vessels out in the storm that night sighted the *Portland* as it labored against the mounting seas. Conditions at sea deteriorated quickly as the steamship left the shelter of Cape Ann on the north shore of Massachusetts and ran into crushing waves measured at over thirty feet high. Cape Ann is located about thirty miles northeast of Boston and is the northernmost point along the Massachusetts Bay. It includes the city of Gloucester and other smaller towns.

Jason Collins, the captain of the steamship *Kennebec*, had left Boston Harbor bound for his home port of Gardiner, Maine, when, fearful of the reported rough waters and pending storm, he returned to the safety of the Boston port. Collins anchored his ship in safe harbor between Boston's Long and Deer Islands. Close to 7:30 p.m., while his steamship and its passengers waited, nestled safely between the two islands, Collins reported seeing the *Portland* steam steadfastly by on its way out to open sea. Collins signaled to the passing ship using a blast from his ship's whistle. Collins later reported that the whistle was intended as a warning to the *Portland* about the rough waters ahead, but there was no response by anyone on board the ship.

Wesley Pingree, the lighthouse keeper on Deer Island in Boston Harbor also reported sighting the *Portland* passing that evening. He later reported that there was nothing that indicated anything was wrong with the ship.

Charles Martell, captain of the tug boat *Channing*, crossed paths with the *Portland* a short distance off Nahant, a small fishing village off the rocky coast of Essex County. According to Martell, "The weather was not bad at the time, but I knew a serious storm was coming." Martell reported seeing several of the *Portland*'s crew on the upper deck. They were waving Martell off. Martell yelled back across the bow of his ship to the *Portland* crew.

"You'd better stop that hollering, because I don't think you'll be this smart tomorrow morning," Martell shouted. Little did he know how tragically foreboding his comments were. Martell gave three short blasts on *Channing*'s steam whistle and the *Portland* responded in kind returning with three short whistle blasts.

Captain Lynes Hathaway, who was stationed at the Thacher's Island Lighthouse, reported seeing the lights of the *Portland* as it continued up the coast. He too reported nothing unusual about the sighting. He, like Captain Thomas, also reported that the weather conditions at the time of their sightings were fine, nothing that might make either of them think that the *Portland* would not be able to outrun the pending storm.

Captain William Scott of the schooner *Windward* also reported seeing the *Portland* off the Thacher's Island coast, and according to Scott, he passed close enough to the steamship to see people standing on deck, waving. According to Scott, "The sea was very calm, but we knew we had a bad night ahead of us."

At 11 p.m., the captain of the schooner *Grayling*, Reuben Cameron, reported sighting the *Portland* approximately a dozen miles southeast of Thacher's Island near Gloucester, Massachusetts. According to Cameron, the *Portland* appeared to have changed its course and was heading for a sheltered port like most ships along the coast had wisely decided to do. Cameron reported that although the storm had yet to strike, the *Portland* seemed to be pitching badly in the rough waters. It was the last time anyone ever reported seeing the ship. At close to midnight, the storm hit, bearing down on the New England coast with all its fury.

Captain Bill Thomas, of the fishing schooner, *Maud S.*, reported sighting the *Portland* that evening sometime between nine and ten o'clock off the south coast of Thacher's Island on the eastern most point of the Cape Ann peninsula. Thomas had a more than passing interest in the *Portland*, since his wife, Hope, was scheduled to return home to Maine aboard it. Although worried about the coming storm and the safety of his wife, Thomas noted nothing unusual about the *Portland* other than it appeared to be running closer to shore than usual. He reportedly told his crew, "There goes the *Portland*. She will probably run close to Thacher's." Thomas had no reason to think that Blanchard was "foolhardy in continuing his journey."

Almost everyone connected with the steamship company, given the growing concern over the treacherous weather conditions, believed Blanchard surely had headed for safety in some port along the way back to Maine. But there was no way to know for certain. Communication on ships like the Portland were limited. There were no telegraph, radio, or telephone communications. Ships had to rely on blasts from their steam whistles, flags, and flares as distress signal.

3: The *Addie E. Snow*

The schooner *Addie E. Snow* out of Rockland, Maine, loaded with a full cargo of granite and bound for Boston, left Portland Harbor on the night of Saturday, November 26, 1898, with a crew of sixteen men. The schooner had been making good time heading to Boston when it ran afoul of the great winter storm. The captain and crew had all they could do just to right the heavily loaded ship, as she was pitched helplessly on the wild and stormy ocean. Despite their best efforts, the crew was unable to keep the ship from being blown southward off its course, heading straight into the seething Cape Cod waters. The same storm-tossed waves also sent the *Portland* reeling along a similar trajectory toward the Cape Cod coast.

4: A Knock at the Door

Grace Pendleton was seated on a couch in the parlor in front of the glowing fireplace, wrapping Christmas presents, when she heard the knock at the front door. It was late, almost eight o'clock. She wasn't expecting anyone. Her husband was upstairs working. She went to the front door, unlocked it, and flung it open.

"Hope!" she cried, seeing her sister standing in the doorway. She was holding onto her two bags full of wrapped presents, snow was gathered on her shoulders, and her scarf was tied over her bonnet to keep it from blowing away in the wind.

"I missed the boat," Hope Thomas said, and stepped inside.

5: Pyrrie

> They were driven back by storme of winde and pyrries of the sea.
>
> –Plutarch, 46 AD–120 AD

The storm had reached hurricane intensity, striking the New England coastline during the night and ultimately leaving in its wake destruction, human suffering, and a record number of deaths. As the great winter gale swept up the Atlantic coast, it produced some of the wildest weather ever seen, combining the effects of a blizzard, a hurricane, and a flood.

The monstrous storm had gusts of up to 110 miles per hour. Temperatures dropped to below zero. There were sixty-foot sea swells. It paralyzed dozens of communities from New York to Maine. The elbow of Cape Cod was ravaged by its force. Some places did not fully recover for weeks.

Those on both land and sea were ill-prepared for the terrible force of the winds. Ships of all makes and sizes made their best attempts to reach the safety of protected harbors. Many were not successful as the storm tore across the Massachusetts and Cape Cod Bays. Large ships at sea that could not make it into a safe harbor scrambled far offshore, trying to ride out the storm. When the storm finally did subside the next morning, miles of the New England coast from Long Island Sound to the farther Maine coastline, were scattered with the debris of shipwrecked vessels.

Splintered and broken masts jutted out from broken and twisted hulls, some thrown far inland, smashed against rocks or piers, some thrown up onto the sandy dunes, others broken in two, shattered, resting in the murky remains below water, their torn sails and damaged rigging and cargo floating out to sea or crashing onto shore. Even those ships lucky enough to have made it back into port before the storm reached its maximum intensity did not escape its fury. Many were smashed against wharfs and pilings, while still others were dragged far out to sea by the tumultuous currents.

6: Up and Down the Coast

In Rhode Island, more than fifty ships were reported damaged or wrecked. At Block Island, off the coast of Point Judith, Rhode Island, winds reportedly reached over 100 miles an hour. More than 30 seamen had been rescued from the storm by the squad of Life Saving crews stationed at Block Island's New Shoreham Life-Saving Station, and the small island's entire fishing fleet was shipwrecked.

The storm moved steadily along the coast hitting Cape Cod and the Islands. More than forty-five ships anchored in Martha's Vineyard's safe harbor were damaged or sunk. In one horrible instance, an unlucky fisherman froze to death, caught in the rigging of his schooner that had been driven onto the island's shoals. Life-Saving Service crews on Martha's Vineyard worked around the clock trying to save those they could. They were not always successful. The violent waves and wind seemingly picked up and threw a two-mast schooner at Gay Head and sent it crashing onto the rocky coast, killing all of the crew on board. None of the islands off Cape Cod, Nantucket, and the Elizabeth Islands, including Cuttyhunk, escaped the wrath of the storm, and neither did Cape Cod itself. A reported fifty ships were wrecked along the Cape Cod shore.

One schooner sank off Provincetown Harbor in full view of the Long Point Lighthouse and Wood End Life-Saving Station. Lifesavers fought to reach the ship's crew, who had climbed into the rigging as the turbulent waters rose up over the ship's decks and cabins, leaving visible only the topmost section of its masts, where the five crew members hung on for dear life. But the surf was too powerful and the lifesavers were unable to reach the crew, who ultimately froze to death clinging desperately to the mast rigging. When their bodies were finally reached, the sailors' hands had to be cut free from the frozen ropes they grasped into death.

CRUSHED BY ANGRY WAVES
Thrilling Experience of the Crew of the Steamer *Strathesk.*
One Man swept Overboard
His Shipmates Powerless to Save Him From Drowning–
The Battered Vessel Makes Port.

The British steamer *Strathesk* came up the bay yesterday morning and anchored off Liberty Island. She was almost a wreck. When buffeting the wind and waves in last Saturday's storm her decks were swept from stem to stern. Every moveable article was carried overboard. The forecastle had been made uninhabitable through big waves washing through it and many members of the crew were suffering from cuts and bruises. This was not the worst. While the gale was at its height a wave swept over the ship and carried away one of the men, Thomas Brewster. His shipmates saw him go over the side of the ship but were powerless to help him. Life Buoys were thrown over with the hope that he might find one but he disappeared.

Capt. Frost, the skipper of the *Strathesk* navigated the vessel 150 miles without a compass and with its steering gear out of order. . . . He said it was the worst storm he had ever experienced and that half of the disasters has not been told yet." (*New York Times*, December 1, 1898)

4: Boston Harbor and Points North

In and around Boston Harbor more than two dozen ships were lost in the storm, some driven far inland during the ferocious gale. Boston Harbor was nearly impassable because of the vast amounts of debris and wreckage left after the storm. According to reports from the day afterward, the two-mast schooners, *Freddie W. Alton* and *Albert H. Harding* were wrecked; the *Alton* lifted from its mooring and carried nearly a hundred yards inland and the *Harding* hurled against the rocky shore and sunk. An English steamship carrying more than five hundred tons of cargo ran aground five miles out of Boston's Long Wharf, near Spectacle Island. It took three days before it could be towed free.

Two schooners headed down from Maine, the *Lizzie Dyas* and the *Virginia* were lost at sea. The *Dyas*, which was carrying a shipment of granite, sunk to the bottom, although its captain and crew managed to save themselves. The crew of the *Virginia* were not so lucky. After being tossed against the rocks just off Thompson's Island in Boston Harbor, the ship sank, taking its entire crew with it.

Farther up the Massachusetts coast, the town of Salem lost nine ships, and all hands on board the schooner *Bessie H. Gross* drowned when it crashed onto the rocky shore. Luckily there were no deaths reported in Gloucester, despite the fact that nearly thirty ships of all types were wrecked, some torn from their moorings and thrown ashore, while others simply sank or were thrown in a hodge-podge of shattered debris on top of one another. In Rockport, Massachusetts, a reported eight small fishing vessels sank trying to ride out the storm in the harbor. The seven-member crew of the three-mast schooner *Charles E. Schmidt* was saved when it ran aground. The captain and crew of the floundering ship stuck fast on rocks along the farthermost Rockport coast, were rescued by a Davis Neck lifesaving crew.

The New Hampshire coast fared better than most, while there was negligible damage recorded in Maine. In Portland, several schooners were run ashore with broken masts, rigging lost, but most was minor damage.

8: New York Devastated

In New York City, conditions were reportedly the worst since the Blizzard of 1888–a storm that packed almost as powerful a punch. In New York, the gale began at noon on Saturday as a light dusting of snow, followed by sleet and rain. By Sunday night the wind had reached fifty to sixty-five miles per hour and a full ten inches of snow had fallen, accumulating into gigantic mounds. In some places, the snow drifts reached over six feet high. As quickly as New York City street department crews could dig through the drifts, the giant mounds reappeared. The hurricane velocity winds and torrential sleet and rain forced the total suspension of all steam and electric car service in and around the city. Most people were stranded, if lucky, in their homes or offices, while others found refuge from the storm wherever and however they could. The tracks of the New York, New Haven, and Hartford and Long Island railroads were invisible, blanketed with heavy snow. In some places along the railroad route, mounds of snow reached over ten feet high. Outside the city, the situation was pretty much the same.

On the farthest tip of the south shore of Long Island Sound, a record-breaking sixteen inches of snow had fallen. Amounts well over a foot were common in the neighboring towns and villages. North of the sound was hit the hardest.

9: Connecticut Buried

Southern and central Connecticut were buried under an average of two to three feet of snow, steadily pounded by torrential sleet and rain, and blasted by hurricane-force winds. All rail, coach, and carriage transportation in and out of the city came to a standstill. In Middletown, Connecticut, nearly forty inches of snow had accumulated over the storm's two-day strike. In places like Bridgeport, Hartford, New London, and Waterbury, snowfall reached into the thirty-inches range. The winds inland were not as severe, but overwhelming nonetheless, reaching a velocity on average of forty-five to fifty miles per hour.

10: Rhode Island Hit Hard

Heading down the coast to Rhode Island, near blizzard conditions left almost two feet of snow in some towns. Bristol reached twenty inches, Providence more than twenty inches, with snow drifts as high as twenty-five feet, which kept the trains from running and the major arteries blocked. For some trains, the storm had struck too fast and too hard, stranding them in mid-route. The New York, New Haven, and Hartford train, heading out of Providence and bound for New York City, was stranded on the tracks with more than one hundred passengers for over twenty-four hours. An express train heading out of Boston and bound for Providence got stuck in an immense snow bank only a mile and a half from the Providence terminal. Some of the more hearty and impatient passengers made the decision to trudge the rest of the way into Providence, where they merely found themselves stranded in the Providence train station. It took train workers more than twelve hours to shovel and plow the tracks enough for the stranded train to finally make its way into the Providence station.

On Block Island, ten miles off the Rhode Island coast, the wind velocity reached an unheard of 90 mph. Gusts were later estimated as high as 110. Huge fifteen-foot waves smashed over the island's breakwater, flooding the island's small downtown boardwalk. Gigantic waves breached Ocean Avenue on Block Island, sending water pouring into the lobbies of several island resort hotels and destroying much of the avenue itself. Most of the macadam roadway was washed away, and most of the hotels on the island sustained major damage. Hotels like the Ocean View, the Eureka, Spring House, and Highland House lost portions of their roofs, and siding. Lobbies were flooded. Cupolas were torn off by the howling winds. Sections of huge porches, like the one that wrapped around Spring House, were nearly lifted from their foundations.

11: Cape Cod Suffers the Worst of It

Cape Cod suffered the worst of the storm. One end of the Cape to the other was devastated. At the Cape's farthest tip, Provincetown, more than twenty buildings, including small fishing shacks, cottages, and businesses were blown down. Nine wharves along the Provincetown waterfront were demolished.

The wind at Woods Hole, near Falmouth, reached nearly eighty miles per hour, setting an all-time record that had been previously recorded at seventy miles per hour during the infamous blizzard of 1888. In the beginning of the storm, most of the downpour that hit the Cape Cod coast and inland came in the form of icy rain and sleet, but by Sunday night the snow began falling in great sheets, covering everything in sight. The town of Hyannis was covered in nearly a half foot of snow.

The hammering rain and flooding washed out sections of the Cape Cod railway line, stranding hundreds of people and temporarily at least shutting down the Cape's lifeline to the mainland. The Cape Cod Canal had yet to be dug. Violent winds blew down telephone and telegraph poles, tearing down miles of communication lines, shutting off any communication the Cape might have had with the outside world. It was not until late Tuesday that communication lines were reestablished.

> The heft of the storm seems to have been about the time or shortly after the center passed over the Cape, which is generally agreed to have been about 9:30 on Sunday morning. The sky at that time over the stretch between Chatham and Barnstable cleared off entirely and the wind died out. Fifteen minutes after, it was blowing hard from the north, and it was at this time that the gale wrought the greatest destruction among the trees from Yarmouth

> to Middleboro. In this respect, Sandwich seems to have suffered the most, for not only did the silver oaks, as they are called, go down, but great elms in the town of Sandwich were blown across the streets, and it was a day or two before the main street was passable. (*Boston Herald*, December 3, 1898)

12: Massachusetts Hit with Indescribable Damage

In Boston, nearly forty people were killed by some extension of the ferocious gale. Sections of the city's once bustling waterfront were left in ruin. Much of the warehoused merchandise that had been stockpiled along the Boston waterfront was destroyed or swept out to sea by the torrential tides that pounded the Boston wharves. Talking to a *Boston Herald* reporter, a local constable claimed that it was one of the worst storms he had encountered in the last thirty years.

"I never saw anything like it in my life. There are coasting vessels ashore in a dozen different places, and it will be necessary to lighten these vessels or dig a channel to them," the man claimed.

Plymouth, Massachusetts, up to Scituate, south of Boston, suffered indescribable damage to homes and businesses along the waterfront. Hundreds of oceanfront houses, cottages, hotels, and other commercial properties were torn to shreds by the ferocious coastal winds and record high tides. Some homes were literally lifted from their foundations by the pounding surf, carried by the hurricane and dropped half-way across town in a twisted pile of broken timber. A dozen cottages located along Plymouth Beach were razed and the bridge crossing Town Brook was lifted from its pilings and carried a half mile away. The belfry in Town Square Church was blown off. The building that housed the Plymouth Yacht Club was rendered a pile of rubble turned over on its side.

In coastal Salem, close to two feet of snow piled up, with winds felling dozens of trees, clogging roads, severing power lines, and cutting off transportation and communication in the city and beyond.

The windows in close to fifty homes in Newburyport were blown out, and dozens of brick chimneys toppled, tossing bricks and debris throughout the town. In Gloucester the railway storage building was destroyed.

The lifesaving station in North Scituate was nearly washed away by the relentless tide and gale winds. The first floor of the station was submerged and the crew of the station had to seek refuge on the second floor and wait until the tides subsided before resuming their duties along the Scituate beaches. More than sixty cottages and homes in the town were destroyed by the turbulent sea.

In the Town of Cohasset, the storm lifted ships out of the harbor, slamming them into the middle of the town's main thoroughfare. Every road in Hingham was flooded, and the steamship pier along the harbor was toppled. More than five hundred feet of Hingham's Summer Street seawall was washed into the ocean. Wagons, carriages, and even rail cars were swept into the streets. Trees and telephone and telegraph lines were toppled, and roads were impassable because of all the debris and mounds of snow. The regular railway service of the New York, New Haven, and Hartford line was suspended because all along the route sections of the ties and bridge supports had been washed away. Snow piled up to a foot and more throughout Boston. Wind velocity hit over seventy miles per hour, creating huge snow drifts that blocked streets and railroad lines, shutting down almost all transportation except for those hardy souls who maneuvered through the drifts on foot.

The storm swept across all of Massachusetts, leaving in its wake a terrible swath of destruction, knocking out power, communication, and transportation from Springfield, where a record twenty-three inches of snow piled up, with some drifts reaching over ten feet tall, to Fall River, where flooding along the Taunton River and Mount Hope Bay washed away dozens of homes and businesses.

The November gale of 1898 is still cited as one of the worst in New England history.

13: Down to the Sea in Ships

The results of the horrific storm were felt up and down the New England seacoast, burying much of New England in an avalanche of snow and ice and destroying hundreds of coastal homes. Newspapers reported approximately 155 ships destroyed in the storm and nearly five hundred lives lost. The U.S. Life-Saving Service reported saving more than 130 people over the course of the devastating storm.

Slowly, after the storm, the sea gave up its dead. Casualties continued to mount as the frozen bloated bodies began to wash ashore among the wreckage. For weeks after the storm, the bodies of men, women, and children were swept ashore in the debris-laden waves, some washed up onto the dunes, while others were dragged from the sea by rescuers.

> I went down the beach to the key post about three miles from the station. . . . It was blowing so hard that I was obliged to kneel down at times to get my breath. . . . I could not see any distance offshore. . . . I warned a number of people in houses nearby that they had better seek safety elsewhere, as the seas were breaking up against the windows. I helped two families, the women and children–to a safe place in another house. (Richard Tobin, North Scituate Life-Saving Station, November 28, 1898)

The four-mast schooner, *Abel E. Babcock*, an 800-ton ship loaded with a cargo of coal, attempted to ride out the storm anchored just off Boston Light. The ship was unable to hold its position in the heavy winds and it was blown up the coast, landing in Hull, where it was smashed against the rocky shore. All eight members of its crew were drowned.

The seven crew members of the schooner *Henry R. Tilton* were more fortunate. During the height of the storm, the ship, loaded down with a cargo of lumber, ran aground several hundred yards north of the Hull coast. The waves were too high and the wind too powerful for the Point Allerton lifesaving crews to dispatch a boat. Instead, the lifesaving crew fired a line from shore to the stranded ship and readied a breeches buoy, which is a rescue apparatus with a leg harness attached to haul people out from wrecked vessels, one at a time. The Point Allerton lifesaving crew had to brace themselves along a sea wall in order to pull the crew members to safety one at a time. Waves and wind whipped over the wall, knocking the lifesavers to the ground and tangling the breeches buoy lines. But they persevered despite the dangers, managing to pull all seven members of the stranded *Tilton* to safety.

The schooner *Calvin F. Baker*, loaded down with a cargo of coal, had been swept across Massachusetts Bay, its mast broken and its sails torn away. It was stranded off Little Brewster Island, a rocky coast island, nine miles off shore. One of the ship's crew tied a rope around his waist and tried to swim to shore to get help. He drowned in the icy waves and his body was never recovered. The remaining crew secured themselves up in the ship's rigging in an attempt to save themselves from the relentless waves that pounded over the stranded vessel. After several hours and nearly freezing to death in the rigging, they climbed back down to the deck where they found shelter in a small cabin located on the ship's forward deck. They stayed huddled in the small shelter as the sea continued to smash against and over the ship. Their clothes, hands, face and feet were drenched and frozen. By nightfall, there were still no rescuers in sight. One of the crew was swept out to sea, unable to hang on to anything, his hands frozen stiff. Another froze to death in the same position he had been huddled in. Finally, a crew from the Point Allerton Life-Saving Station made their way to the stranded ship. Launching a surfboat, the lifesaving crew rowed out to the ship, which was nearly broken in two and covered in ice. They climbed on board and found the five remaining crew members still huddled together in their small forward deck shelter. They were unable to move and had to be carried to the waiting surfboat one by one. On shore, the men's clothes, boots and gloves had to be cut away they were frozen so hard. It was days before the five men were released from a doctor's care.

ECHOES OF THE GREAT GALE

Vessels Limping Into Port After Battling
with Wind and Waves–
Accidents and Wrecks Reported

> Every vessel that reached port yesterday had experienced the hurricane that blew last Saturday and Sunday, and the Captains and crews all told thrilling stories. All agreed that the storm was one of the worst known, and many said that it will be weeks before full reports of the death and disaster at sea are in, and it may be that some vessels have disappeared that will never be beard of again. (*New York Times*, December 1, 1898)

The results of the horrific storm were felt up and down the New England seacoast, burying much of the region in an avalanche of snow and ice and destroying hundreds of coastal homes, leaving four hundred dead and more than two hundred ships damaged.

> LOSSES OF MARINE UNDERWRITERS
> Local Managers Estimate Them at
> Over $1,000,000.
>
> The marine underwriters of this city say that the losses sustained by their companies as a result of the storm beginning last Saturday will probably amount in the aggregate to something more than $1,000,000. Owing to the lack of sufficient data of a trustworthy character, it will be some days before a definite statement of the full extent of the losses can be made. (*The Boston Herald*, December 2, 1898)

14: The Weather Outside Is Frightful

A compendium of storms that have ravaged New England is far too lengthy to catalog here. Suffice to say, from the shores of Connecticut and Rhode Island, and along the beaches of Massachusetts, New Hampshire, and Maine, storms, blizzards, and hurricanes were the bane of the New England maritime industry, causing great suffering and chaos. Thousands of lives, homes, buildings, ships, and businesses have been lost over the years. Whole beaches have been washed away.

The proximity of New England to the ocean was and is both a blessing and a curse in terms of calamitous weather. It was, in fact, a New England storm that blew the Pilgrims off course forcing them to land in Plymouth. The "Portland Gale of 1898," so named after the steamship *Portland*, was so powerful, it formed a new route for the North River in Scituate, Massachusetts, which separated the Humarock Peninsula that was previously attached to the town. Some storms leading up to the Portland Gale are worth noting.

> Before 1898 the mouth of the North River sat at the southern end of Humarock, near present-day Rexhame Beach. The site of the current mouth between Third and Fourth Cliffs was then a narrow barrier beach. The winds and extremely high tides of the Gale stirred up the waters of the North River with such vigor that by the storm's end the river had washed over the beach, cutting itself a new outlet to the ocean. The old mouth, two miles away soon filled in with sand. (Kezia Bacon, "The Storm That Changed the River's Flow," *Mariner Newspapers*, 2008)

15: The Hurricane of 1778

The hurricane of 1778, known as the "Revolution Storm," hit New England in the midst of the War of Independence. It struck Cape Cod, Massachusetts, killing between fifty and seventy people, twenty-three of them attributed to the sinking of the British warship, HMS *Somerset III*, which sank on the shallow Peaked Hill Bars, just off Race Point Beach in Provincetown, Massachusetts.

The Cape Cod coast was hit with torrential rain and snow and high winds and tides. The *Somerset* was trying to return to a safe harbor in the middle of the howling gale. It ran aground on Provincetown's outer bar, where it foundered.

Initially citizens lined the beach in an effort to save the lives of her crew, even though they were their enemies. On board the ship, boats were launched, but they were dashed to pieces. The British sailors drowned. Guns, shot, and other heavy articles were thrown overboard, her masts, which had been broken off at the deck, were cut adrift, and finally, at high water, the leaking hulk was driven, by the force of the wind and sea, over the bar and up on the shore. Only twenty-one of the four hundred British sailors on board survived. The survivors were greeted by angry Provincetown residents, who made sure the sailors eventually made their way to prison in Providence, Rhode Island.

16: The Storm of October 1804

The 1804 storm brought heavy snow across the Northeast, in some areas up to two to three feet. It killed nine people and caused a reported $100,000 in damage across the region. It became known as "Snowcaine," because of the vast amount of snow it brought with it as well as its hurricane-force gales. In Boston Harbor, several ships were wrecked. The storm produced record-breaking gusts, rainfall, and snow. In Massachusetts, between five and fourteen inches of snow were reported.

In Salem, Massachusetts, seven inches of rain fell, the largest amount ever recorded during a single twenty-four-hour period. In Concord, New Hampshire, two feet of snow was recorded. Western New Hampshire along the Connecticut River was blanketed with eighteen inches of snow. Homes, barns, sheds, and chimneys did not escape the storm's wrath. Roofs were blown off, houses collapsed, and many people lost their lives. Streets were filled with the debris of destroyed buildings, fences, and trees.

Many ships in Boston Harbor and elsewhere lost their moorings and were wrecked, smashing into one another or being driven ashore and dashed against the wharves. Some ships were hurled ashore or against reefs and sand bars. In Cohasset, Massachusetts, the sloop *Hannah* was blown out of Cape Ann Harbor and struck a ledge off shore, miles away. But *The Hannah* was one of five vessels wrecked on the shores of Cohasset during the storm.

17: The Storms of 1841 and 1842

The October Gale of 1841 dropped up to eighteen inches of snow and sleet in Connecticut and some places in Massachusetts. It wrecked the Georges Bank fishing fleet and approximately eighty-one fishermen drowned. It knocked down trees, tore roofs off houses, and forced boats up on shore. The storm also destroyed a saltworks factory along Cape Cod, sending the economy into a slump. The gale was so powerful it tore the newest and strongest canvas sails into shreds and toppled masts. Ships anchored in harbors along the coast broke away from their moorings and collided or were smashed against wharves. The tide rose so high that some wharves were submerged.

A ship from Bath, Maine, was driven out of the harbor and carried into Massachusetts Bay where it hit Cohasset rocks and went ashore wrecking on the Scituate beach. There were seven passengers on board–four women and three men–and four crew members including the captain of the vessel. The captain, his daughter, and five passengers including all the women and one man and one seaman perished. At Cape Ann, fishermen lost fourteen of their fleet of sixteen vessels. Many fish-processing houses and about sixty barrels of mackerel and three hundred empty barrels were destroyed at an estimated cost of $50,000. On the Island of Nantucket, several ships tied up along the dock were driven ashore as the tide rose two or three feet above the wharves and ran into the town, flooding streets.

The greatest loss of life and property occurred on Cape Cod. The beach from Chatham to the highlands was literally strewn with parts of wrecks. Between forty and fifty vessels went ashore on the sands there, and fifty dead bodies were recovered. The schooner *Bride*, in Dennis, was driven ashore and the bodies of the eight crew were found in the cabin. Dennis lost twenty-six young men in total. Another of the Truro fleet, the *Prince Albert*, went down on the shoals of Nantucket, and the crew of eight, all of whom lived in Truro, perished.

The schooner *Ellis* of Plymouth, Massachusetts, went ashore on the east side of Truro, and its crew of eight were all drowned. Most of the Truro fishing fleet was on or near the southwest part of George's

Banks before the storm struck and tried to make it safely back to harbor. The violent currents carried them off course and they were driven upon the Nantucket shoals. These unfortunate fishermen were nearly all young men under thirty years old and fifty-seven from Truro were lost. This was the most serious catastrophe experienced by the small Cape Cod town. In commemoration, the town erected a memorial, a plain marble shaft rising from a brownstone base, which is inscribed with the words: "Sacred to the memory of Fifty-Seven Citizens of Truro who were lost in seven vessels which foundered at sea in the memorable gale of October 3 1841. Then shall the dust return to the earth as it was and the spirit shall return to God who gave it. Man goeth to his long home and the mourners go about the streets."

A year later, on Wednesday afternoon, November 30, 1842, a snow storm hit New England with a vengeance. In some parts, so much snow fell that travel on roads and railroads was halted. Fifteen inches of snow was reported in Dover, New Hampshire. The temperature in Belfast, Maine, was recorded at 6 degrees above zero, the coldest November day that had been known there for several years. In Boston the storm was much more severe than at any other port. Many ships were anchored in the harbor when the storm hit and they were driven from their moorings by high winds and rising tides, jammed against each other or the wharves. Several of them were sunk. In the heart of the city the sounds of falling masts and of vessels crashing together were heard from time to time above the noise of the storm.

Among the many wrecks caused by the storm, one of them was the sinking of the *Isadore*, a new, 400-ton ship owned by Captain Leander Foss. This was its first trip, and it sailed on the morning of the storm from Kennebunk, Maine, headed for New Orleans. In the blinding snow and pounding waves, the ship was driven onto rocks near Cape Neddick, Maine. The entire crew of fifteen, all from Kennebunkport, Maine, were lost. Five were fathers of families, and they left twenty children among them. Two were young men, the only sons of widows.

Perhaps the most tragic wreck caused by the storm was the sinking of the schooner *Clark*. It was on a trip from St. John, New

The *Portland* on one of its many trips between Boston and Maine
Samuel Ward Stanton
Antonio Jaconsen

Launch of the *Portland*
Public domain

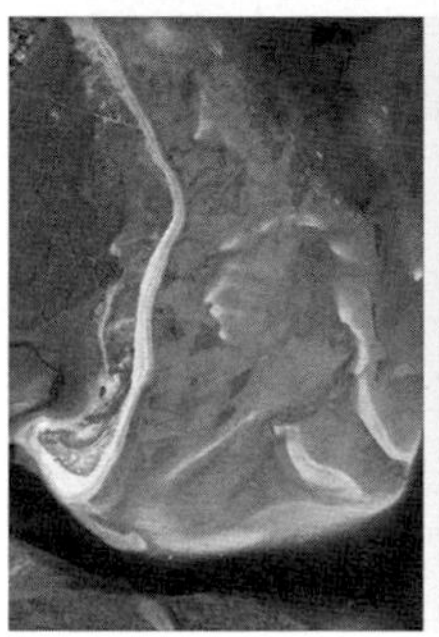

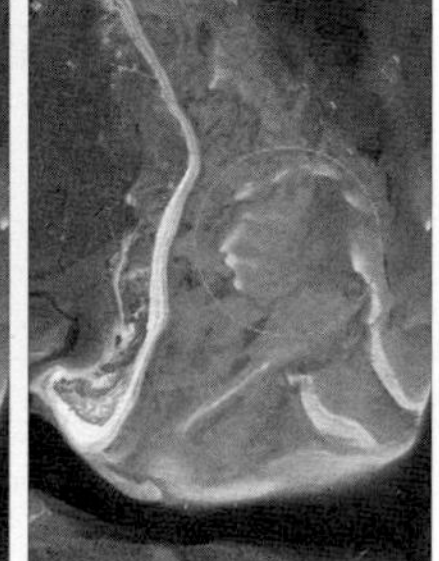

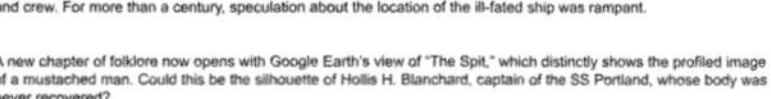

Captain Hollis H. Blanchard
Public domain

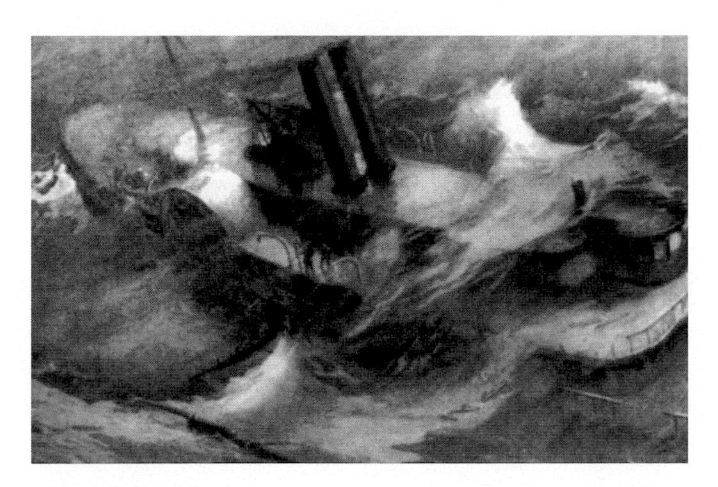

The *Portland* in the storm
Public domain

Portland's boiler steam pipe.
NOAA/SBNMS, 2002

Below the sea: Dishes spilled in the galley
Bob Foster

Port of Portland, Maine

Account of Wages and Effects of a Deceased Seaman.

Name of Ship.	Official Number.	Port of Registry.	Name of Master.	Description of Voyage or Employment.
Steamer Portland	150488	Portland Maine	H. H. Blanchard	Boston to Portland

Name of Seaman.	Date of Engagem't	Time of Death.	Place of Death.	Cause of Death.
Arthur A. Johnson	Nov. 1/98	Nov. 27/98	Near Cape Cod	[illegible]

If the Seaman's name is not on the articles, in this space must be entered the date of his being sent on board the ship; and in such case, here state by whom and where he was sent on board, AND ANY OTHER PARTICULARS.

Wages, Money, Clothes, and other Effects.	Amount.	Inventory copied from the Official Log Book of Articles sold and the sum received for each.	Amount for each Article sold.	DIRECTIONS. Particulars of Deductions.	Amount.	Initials of Shipping Commissioner against each Item checked.
Wages at $22.00 per Month	$22.00					
Money in possession of the deceased	—					
Proceeds of Sale of Clothes and Effects, as per account	—					
Total	$22.00	Total		Total		
Deductions as per Account	—					
Net Amount	$22.00	For Inventory of Articles unsold, see Endorsement.				

I HEREBY declare that the above is a true and correct account of the Money, Wages and Effects of the above-named Deceased Seaman.

Dated this 19 day Dec 1898

Lost — Mate.

Lost — Seaman.

Signature of the Master of the Vessel, Agent — J. F. Liscomb

The particulars respecting the deceased in the following columns are to be filled up IF, and SO FAR AS, THE MASTER IS ACQUAINTED WITH THEM; and if any Will has been deposited with the Master, it must be given to the Shipping Commissioner.

Birth Place.	Age.	If any Will has been made, Name & Address of Executor	Married or Single	If married, the Name & Residence of his Wife.	If any Children their Names and Ages.	Name and Residence of Father and Mother or of the nearest known Relations.
			Yes	761 [illegible] St. Portland		

I HEREBY certify that I have examined the above Account, and compared the Inventory with the Official Log-Book, which is attested by the Mate and one of the Crew; and that the above is a true copy thereof. The Balance of the Account has been paid, and the unsold articles have been delivered to me.

19 day of Dec 1898

(SEE ENDORSEMENT.)

Geo. Tolman, U. S. Shipping Commissioner.

Wage papers filed with the Circuit District Court, 1898.

Public domain

2 PORTLAND DAILY ADVERTISER, NOVEMBER 29 18

PORTLAND A TOTAL WRECK.

All on Board Perished So Far as Known.

LOST OFF CAPE COD SUNDAY.

Thirty Four Bodies Recovered---The First News Came to Beston by Train.

Boston, Nov 29—A special to the Herald from North Truro says the steamer Portland of the Boston and Portland steamship company line, plying between Boston and Portland, was totally wrecked at 10 o'clock Sunday morning off Highland Light.

The crew and passengers perished within a short distance of land.

A large quantity of wreckage including trunks and other material has come ashore, and at dark last night 34 bodies had been recovered from the surf by the Life Saving crew at the Highland saving station.

The news of the disaster was brought to the Herald through the means of a special train, as communication by wire to points on Cape Cod is impossible on account of the havoc wrought by the storm.

The first body was found by Gideon Bonley, a surfman at the station, who at once gave the alarm. Then a thorough search was made.

The bodies saved were evidently those of passengers.

The body of a woman was well dressed.

It is thought that the Portland took the storm outside Saturday night, causing her to break down and finally to drift on the lee shores and to destruction.

WHY THE PORTLAND STARTED.

It Was Contrary to the Direct Orders of Mr. Liscomb Given In Boston.

Newspaper account of *Portland* wreck

NOT ONE SAVED.

The Steamer Portland Lost With All On Board Sunday Morning.

Driven Clear Across Massachusetts Bay by the Fierce Gale.

She Struck on Peaked Hill Bar and Went to Pieces.

The Bodies of the Victims Washed Ashore Near Provincetown.

Many Portland People Were on Board the Ill Fated Craft.

Newspaper account : Not one saved
Boston Herald

Brunswick, to Boston with twenty people, including passengers and crew, on board. The ship left Portland, Maine, on the morning of the storm and late that afternoon was driven ashore at Rye Beach. As the ship foundered, the cabin where the crew was huddled caved in, and the people had to remain on deck. Cold and violent waves washed over them, and they were tossed from one side of the vessel to the other. They suffered intensely from the exposure to cold and water. Some died, the first being passenger Margaret Stewart's six-month-old baby boy, who died in her arms. The mother became insensible and when rescued was found among some lumber almost covered with water. Her arms were stiffened in the position in which she had held her child and remained so for some time after being taken ashore.

Also on board the ill-fated vessel was Mary Hebersen, a widow of about fifty years of age who was accompanied by her daughter, Hannah, who was twelve years old. For hours they managed to keep together in their hopeless condition. But eventually, the daughter, who had become frozen, lay down on the deck at her mother's feet and died. While she lay there, her mother watched over her, helpless to do anything to save the little girl. The mother lay down beside her daughter covering her with her own body and died clutching the frozen lifeless body of Hannah.

One of the sailors on board finally, despite great odds, took a long rope, fastened one end of it on the deck and jumped into the raging surf with the other end tied to him. He fought his way to the shore and by means of the rope, the captain and crew and ten of the passengers were saved.

18: The Great Blizzard of 1888

A decade before the Portland Gale, New England was hit by the Great Blizzard of 1888, often called "The Great White Hurricane." It was one of the most severe blizzards in the history of the country. Snowfalls of forty to fifty inches fell in parts of New

Jersey, Massachusetts, and Connecticut, and sustained winds of more than one hundred miles per hour produced snowdrifts in excess of fifty feet. It was a benchmark for all other New England storms. Telegraph lines were disabled, and Boston was isolated for days. Over two hundred ships were grounded or wrecked, and nearly one hundred seamen lost their lives. It paralyzed the entire East Coast, disabling telegraph communications for days. After the storm, New York began placing its telegraph and telephone cables underground to prevent their destruction. Two women working at a Bridgeport, Connecticut, factory decided to head for home in the storm rather than compromising their reputations by sleeping overnight in the factory with the men who worked there. They were found dead in each other's arms the next day when the digging-out began.

19: Between the Devil and the Deep Blue Sea

> I, with my partie, did lie on our poste, as betwixt the devill and the deep sea.
>
> –Robert Monro, 1637

Four hours out, the *Portland* was nowhere near its destination. Blanchard tried to stay ahead of the storm, reaching Thacher's Island off the tip of Cape Ann, but the storm, with its near 100 mph winds and sixty-foot swells, freezing sleet, and pounding snow, had driven the ship far off course.

The *Portland* was never built to withstand a storm of this magnitude. The gigantic swells hoisted the ship's bow out of the water. Wave after wave hit the helpless ship, lifting its starboard paddle wheel out of the water and then slamming the whole ship back down again. The *Portland* was wide, shallow, and top-heavy. Her design was more appropriate for sailing up rivers than for open sea voyages. Writing for *Scribner's Magazine*, Sylvester Baxter asserted:

> When the coastwise packet steamboat *Portland* was reported missing it was hoped that she might be heard from safe in the open sea, but knowing-ones felt that a steamer of her type had little chance for safety in that storm, for she was of the same general pattern as the Long Island Sound boats—a side-wheeler, with a deckhouse in three tiers. While she was staunch enough for ordinary bad weather, only a propeller of the ocean-going type could be suitable to a route like that between Boston and Portland, all the way through the open sea. (Sylvester Baxter, *Scribner's Magazine*, November 1899)

The design of these coastal steamers allowed them to connect Boston with Maine's river ports. However, it also made them unseaworthy. The *Portland* measured only ten feet, eight inches below the water-line. Yet her superstructure was tall and, being built of wood rather than iron or steel, heavy. This meant that the *Portland* was prone to roll over in rough weather and high seas.

The shape of her hull did not have a sharp bow to cut through the water like many ocean-going vessels of the period. Even worse, the wooden paddles could easily be smashed by a breaking sea hitting the ship broadside. There are stories on record of ships' crews shifting the cargo of paddle-wheel steamers during a storm so that the paddle on the windward side of the ship would be lifted out of the water. This protected the paddle from rough seas but also meant that the ship would only have the use of one of its paddles. These ships were recognized as being so unseaworthy that after the tragedy of the *Portland*, the Portland Steamship Company replaced all of their side-paddle ships with more seaworthy stern-propeller ships.

20: No Visibility, No Means of Communication

Blanchard's visibility was limited. The relentless sheets of snow and sleet were blinding. He had only a slim chance of making it to a safe harbor. The waves and wind were too powerful to navigate.

His only chance was to make for the open sea in hopes of riding out the storm there. In the engine room below, his crew worked frantically to shovel coal into the boiler to raise enough steam to keep the ship afloat. Frightened passengers had to seek safety from the storm inside and must have huddled in their cabins as the steamship was tossed violently on the huge, mounting waves.

The ship had no means of communication to shore so there was no way to let anyone know where they were or to issue distress calls. The *Portland* lost headway against the mountainous waves and howling winds. Blanchard must have decided, given the velocity and direction of the wind, to steer southward, letting the wind and waves carry the ship across Massachusetts Bay toward Cape Cod. His thoughts must have been that if he could reach Cape Cod, he might be able to swing around and drive the ship onto a sandbar, where he might stand a better chance against the storm. He knew he had enough coal to ride out the storm if he could reach calmer waters farther out at sea. There were more than sixty-five tons of coal on board, and the usual nine-hour trip from Boston to Portland used only twenty-five tons. If the bulkhead held and there was no structural damage, he could survive the brutal storm.

> Contemporary interviews of knowledgeable mariners suggest that Hollis Blanchard would have kept the steamer headed into the waves, hoping to outlast the storm. For an as yet undetermined reason, the *Portland*'s bow swung to the southeast, putting the steamer broadside to the waves piling into Massachusetts Bay from the northeast. ROV [remotely operated vehicle] video revealed that the *Portland*'s rudder is in place and turned to starboard, possibly indicating that Captain Blanchard initiated the steamship's turn to the south near the site of its loss. The waves and screaming wind breached the superstructure allowing water to fill the hull. (Mathew Lawrence, Deborah Marx, and John Galluzzo, *Shipwrecks of Stellwagen Bank: Disasters in New England's National Marine Sanctuary*, 2015)

Considering the design of the ship and that it had not been built for the open sea, Blanchard's skill as a seaman in managing to keep the ship afloat during the violent gale was almost miraculous. Turning the *Portland* into the storm propelled the steamship along at the speed of a landslide careening down the steepest mountain. Along the port and aft decks everything that had not been secured, and even the few things that had been, were ripped free and tossed overboard. In the bar and kitchen, glasses, dishes, pans, pots, silverware, and plates crashed to the floor. Mountainous waves swept over the railing, scattering coal bins on deck and hurtling the bits of coal like projectiles, hitting doors and hatches like a fuselage of dark bullets. Flashes of lightning and exploding thunder broke overhead.

Blanchard had already ordered a distress flag hoisted into the rigging, but there was no one, no other ship out in this storm, to recognize the signal flying from the mast. Ships all along the coast had long ago raced for shelter. Those that hadn't had probably already wrecked. The gigantic swells hoisted the battered ship's bow up and out of the water and then sent it crashing back down again.

Blanchard ordered the ship's stewards and crew to begin issuing life preservers. Stewards and cabin boys inching along the deck, hung on for dear life, blown by the raging wind, rocked by the accelerating waves as they crashed again and again against the ship. Blanchard knew it was too dangerous to order passengers and crew into the life boats. They would all be swallowed by the raging sea.

Men, women, and children, huddled in their cabins, some hysterical, others praying, some simply trying to comfort their children, all of them overcome by uncertainty and fear. Some others might have sought comfort, if not bravery, by making their way to the ship's saloon and consuming whatever liquor had not been tossed from the shelves.

21: Margaret Heuston

Back in Portland, high atop Munjoy Hill, Margaret Heuston stood quietly over the mixing bowl on the kitchen table in front of her and waited. In front of the old cast-iron stove was a stack of firewood she had brought inside before the storm. She had filled the bottom of the stove full of wood and it burned so hot the top of the stove became red hot. A pan of freshly made corn bread was cooking. Outside, the howling of the November wind reminded her of singing off in the distance, but she was unable to hear the words, only the haunting sound. The snow began falling in furious sheets, covering everything outside quickly, with mounds building up along the walkway, sidewalk, and street in front of the house. She tried not to think about the storm. She tried not to think about her husband on his way back to Portland from Boston.

She had not heard from him yet, but she knew she would. She always heard from Eben. He was a homing pigeon when it came to his family. She knew that as soon as the *Portland* put in, Eben would throw on his wool winter coat, pull on his galoshes, tug his black wool cap down over his head, and make the long hike up the hill from the Portland pier to their house. He would be hungry for home cooking. He always was. Anticipating his arrival in a few hours, she had made his favorite–Down East corn bread. She made extra, just in case he was real hungry.

Rumors had been flying from Boston to Portland about the intensity of the storm, but no word of the whereabouts of the *Portland* had reached either port. If it grew worse, Margaret Heuston was certain that Captain Blanchard would put in somewhere along the way. She had heard that the *Bay State* was staying in port to ride out the storm. That was the smart thing to do, she thought. Maybe Blanchard decided to do the same in Boston, but there was no way of knowing. All lines of communication were down. She had tried to call the steamship office in Boston to find out the status of the *Portland*, but there was no answer.

Margaret and Eben had only been married a year, but she knew him like a book. She knew more about Eben in the one year they had been married than she had known about her first husband, Ray Ball, in all the five years of their marriage. Ball had died at sea two years before she and Eben were married. His body was never found.

She married Ray while she was still in high school. He was almost twenty years older than she was. He was a quiet, solitary man, set in his ways, as many of the men who fished the sea were–some by nature, others by habit. He smoked and drank, and although he considered himself a Christian, he never attended the Abyssinian Church with her, except on holidays like Christmas and Easter, and of course for the baptism of their two children. He was a good man, kind to her despite his rough ways. Ray Ball could not have been more different than Eben.

Eben was five years younger than she. They met at church and Eben claimed he fell in love with her the first time he laid eyes on her. He courted her for a year, bringing her flowers and candy. He took her out to the fanciest Portland restaurants. He helped her with the small apartment she and her two young children lived in. He sat beside her in church and held her hand as they prayed together or when they stood to sing. They were married within a year.

One of the first things he did for his new bride and stepchildren was to buy a home on Munjoy Hill, just a block away from the church. Now, Margaret Heuston, married just a year ago, waited nervously in that same house for her husband to return safely from the sea, the sea that had already taken her first husband from her and left her alone in the world.

22: Down East Corn Bread Recipe

Ingredients

- 2 cups cornmeal
- ¼ cup sugar
- 2 tablespoons all-purpose flour
- 1 tablespoon baking powder
- 1 teaspoon baking soda
- 1 teaspoon salt
- 2 cups buttermilk
- ¼ cup vegetable oil
- 1 large egg

1. Preheat oven to 425°F. Grease a 9x9-inch metal baking pan.
2. In large bowl, combine first six ingredients. In medium bowl, whisk buttermilk, oil, and egg until blended; stir into cornmeal mixture just until moistened.
3. Pour batter into prepared pan. Bake 20–25 minutes or until golden and toothpick inserted in center comes out clean.
4. Cut cornbread into nine squares.

23: The Boiler Room

Harry Rollinson saw to it that everything was under control in the *Portland*'s boiler room. The fires had been drawn and the ship was building up a full head of steam. He ordered most of the stokers to their stations and the ship's two engineers, Hamstead and Groves, tended to the pumps. The boiler room was clouding up with steam from the huge furnace, causing the crew to take on almost inhuman shapes as they hurried through the engine room. The fire roared and the rhythm of the steam pumps was fast and unfaltering.

Blanchard called down from the bridge for more steam. His voice was flat and without enthusiasm. Rollinson shoveled coal rhythmically, one huge shovelful at a time, his muscles straining. The engineers adjusted the engine values and frantically oiled the pumps. Steam gushed through the boiler room as the crew sweated away at their jobs. Pressure was building. One of the engineers called the bridge and told Blanchard the ship was full-steam ahead. The huge black smokestack in the middle of the ship blew off glowing sparks and black clouds of smoked from the roaring engine below deck.

Eben Heuston moved across the decks from stateroom to stateroom offering assistance, bringing blankets and providing comfort as best he could. George Kenniston Jr. stayed in his luxurious stateroom on the top deck, reading. Emily Cobb was alone on deck clinging to an inside railing. Former State Senator Freeman left his stateroom and headed to the saloon, where they were offering complimentary cigars and drinks. Charlie Thomas and his wife, young daughter, and their dog stayed huddled in their cabin.

24: Elsewhere on the Ship

Blanchard steadied himself against the brass railing in the pilothouse on the upper deck as the ship was slammed by yet another thunderous wave. A collective cry rose from passengers huddled on board. He had all he could do to keep the huge paddle boat righted in the mountainous swells. At times the ship floundered. Blanchard tried to steer southeast, away from shore. The ship rolled and the engines far below growled, as Blanchard tried to say ahead of the thunderous waves. He called below to the engine room and ordered more steam. He needed all the power he could get to keep the ship's bow in the water. Frightened passengers clung to whatever they could to steady themselves. Children cried. Their parents tried to console them. Some families huddled together in their staterooms and prayed.

Thunder claps resonated in the darkness. Luggage, bags, parcels, anything not secured flew dangerously from storage racks. In the kitchen, dishes, pans, pots, and silverware crashed to the deck floor. On deck, what the crew hadn't been able to tie down, from barrels to benches, flew wildly along the upper and lower decks, hitting the railing. Some things were completely lifted by the wind and tossed overboard.

The ship was losing headway against the mountainous waves and relentless wind. As Blanchard tried to come about, the main beam began to give way. A deathly resounding crack reverberated through the ship. Ship stewards and cabin boys inched along the decks with armfuls of life preservers. They hung on for dear life, battered by the torrent of rain, the wind, and the accelerating waves that crashed again and again against the ship. Eben Heuston carried a stack of life preservers along the top deck, stopping at one stateroom after another, helping frightened passengers, men, women, and children, put on the cork-filled preservers.

"Better safe than sorry," Heuston said at each stop. "Nothing to worry about. It's just a precaution. Nobody's better than Captain Blanchard at this. He'll see we all make it home again safe."

Beneath his breath, Heuston quietly prayed that what he said was true.

25: Kenniston and Cobb

There was a loud grinding noise, followed by a sudden jolt. It seemed to come from somewhere far below deck. The jolt caught George Kenniston Jr. off guard and nearly knocked him out of the chair he was sitting in, busily writing in his leather-bound journal. He imagined that this trip would at some later date make the basis of a good story he would someday write. He rose from his chair, steadied himself and made his way to his cabin door. He had to walk with wide, uneven steps to balance. Cautiously, he made his way to one of the stateroom's portals and peered out. For as far as he could see the dark sea was roiling, the white-capped waves smashing hard against the side of the ship, sending streams of spray lashing up and over the guard rails as high as the top deck. The snow was falling like a white curtain.

Just outside his door, he saw Emily Cobb clinging desperately onto one of the inside railings. She was soaked to the bone and could barely stand. Kenniston flung open his stateroom door and called to her. His voice was barely audible over the howling wind. He finally was able to catch her attention. Slowly she inched her way toward the open door where he invited her in. "It's s safer in here," he told her. "And warmer." Still she hesitated. He offered to leave the door to the stateroom open for appearances' sake. Finally, she stepped inside.

Kenniston offered her a towel to dry off. He had a pot of tea that somehow had managed not to crash to the floor and poured her a cup. She drank it quickly.

"Are we going to sink?" she asked her voice barely above a whisper.

He assured her they weren't.

"Captain Blanchard is a good captain. He'll weather this storm," he said.

Cobb sat down on the couch opposite Kenniston. He took up the chair he had been sitting in previously.

"I hope your music hasn't been ruined," he said.

Cobb checked on her sheet music inside her bag. Some sheets were soaked but not completely damaged.

"I know them by heart, anyway," she said.

"Maybe I can stay over in Portland when we get back and come to the recital," he said.

"What are you writing?" she asked pointing to the leather-bound journal that lay on the floor beside him. She read the brass embossed title on the journal: "My Life and Times."

He was embarrassed.

"My mother gave it to me. I'm going to be a writer," he said. I'm keeping a record of everything I see and hear so I can use it someday."

There was another ear-splitting creak and another resounding jolt that shook them both nearly out of their seats. The starboard side of the ship swung around violently. The hull had cracked. Now, it was at the mercy of the waves.

26: Back in Portland

Boothbay banker George Kenniston Sr. waited for some word of his son, George Jr. The weather was getting worse and there was still nothing on the whereabouts of the *Portland*. Kenniston and his wife, Anne, took a carriage from Boothbay to Portland to await word on the missing ship. They were told that because of the storm all the telephone and telegraph lines up and down the coast were down and that there was no way of knowing if, when, or where the *Portland* might have found safe harbor. They were assured, however, that Captain Blanchard surely must have headed into port somewhere when the storm came up. Still, Kenniston feared for his youngest son's safety. He was no stranger to tragedy. He had lost his first wife, Antoinette, young George's mother, in 1881. He remarried shortly thereafter to Anne Blair, two years later.

Judith Cobb, the widowed mother of Emily Cobb, waited in Portland for some word of her daughter. The Reverend John Carroll Perkins, associate pastor at the Unitarian First Parish Church on Congress Street, stayed with her, trying to comfort her during her long ordeal.

Hundreds of others, just like them, men and women, husbands and wives, fathers and mothers, prayed and waited for some word of their missing loved ones. But there was no word coming. For three long, angst-ridden days there was only silence. When word finally did come, it cast a pall over the New England coast, from Boston to Portland.

27: Back in Boston

Back in Boston, Hope Thomas told her sister Grace and her brother-in-law, John Pendleton, how she had been standing in line waiting to board the Portland when she decided to go back and exchange the shoes she had bought. She showed her sister the two-tone shoes she had exchanged the original purchase for.

"I'm so glad you took my advice," Grace laughed.

Hope explained that by the time she exchanged the shoes and made it back to India Wharf, the Portland had sailed.

"Just as well," John Pendleton said. "Who knows if it ran into this storm. We're glad you're safe with us."

Hope said she was worried about her husband, who was still out at sea aboard the *Maud S*. There was no way of contacting him, and she prayed he made it into a safe harbor somewhere.

"He's a smart fisherman," Grace said. "He'll know what to do."

"He still thinks I'm on board the *Portland*," Hope said.

George Gott, the Brookline, Massachusetts, shoe salesman who had decided not to board the *Portland* after seeing the black cat abandoning the ship stayed over in Boston at a hotel. He had planned to take the train to Portland, but even the trains had stopped running because of the storm.

Carrie Courtney who spoke to Captain Blanchard just before the ship left Boston Harbor and to whom Blanchard had shown the watch he bought for his daughter, also spent the night in Boston at a friend's house to wait out the storm.

And Anna Young and her baby who had disembarked from the *Portland* after receiving the telegram from her mother returned home to her husband at their home on Boylston Street, where her husband thanked God that his mother-in-law had sent the telegram.

The *Portland*'s first pilot, Lewis Strout; first mate, Edward Deering; and the clerk to the ship's purser, John Hunt, who had been given permission to attend the funeral of Captain Charles Deering, all stayed in Boston overnight. Although they had planned on returning to Maine on board the *Bay State* the next day, because of the massive storm, it was three days before the *Bay State* was able to make its trip back to Boston.

All these were saved. Others were not so lucky.

28: Down in the Hole

Down in the hole, Harry Rollinson, below deck in the engine room, barked orders to his men in rapid succession, urging them to work harder, dig deeper, and keep the furnace fires blazing.

"Captain needs steam, boys!" he bellowed.

The engine room crew frantically shoveled coal into the blazing boiler, their muscular bodies, glowing like animated silhouettes in the glow of the hot red flames that lapped out of the huge boiler. The boiler roared; sparks flew in every direction.

Up on the bridge, First Mate John MacKey ordered the pilot to pull the engine speed to a complete stop. A nervous minute passed as everyone on the bridge looked to Blanchard.

There were two propellers on the *Portland*. One of them was gone. Blanchard hoped he could still maneuver the ship out into the open sea with just one propeller.

"Back to full steam ahead," he commanded the ship's pilot, Lewis Nelson. He ordered MacKey to check down below in the engine room.

Below, all the water-tight doors were secured. The warning bell sounded above the exit door. The ocean cascaded down the stairwells. An icy foam swirled down the pipes and valves. The chief engineer ordered his men out. The ship was sinking. The crew scrambled madly, leaping through the boiler room door, leaving only Rollinson in the engine room. He yelled after them.

"Secure the door!"

They hesitated.

"Mr. Harry, you come too," one of the last men out called behind him, holding onto the watertight door.

Rollinson waved him off.

"I got business to tend to," he yelled back and picked up another shovelful of coal.

The door slammed shut. A thick jet of sea water poured through the seam in the ship's starboard. He instinctively ran his hand over the gold earring in his ear, before the ship shuddered again. Rollinson lost his footing and had to dig himself out of an avalanche of coal that had poured out of one of the bins. The furnace stokers and firemen, all able-bodied seamen, stumbled in the dark stairwells, picked themselves up, called back and forth to each other, as they hung on for dear life. It was complete chaos.

Several of the crew made their way through the long corridor of the *Portland*'s breached hull to the next compartment. There was frigid water bursting through the seams of the ship's hull. One of the men tugged on the door to the next compartment, opening it wide, and a torrent of water cascaded down the wrought iron staircase overhead, washing over them and sending them flying in every direction. It was no better in the bow, where a foot of water had flooded into the mail room. Stacks of mail floated in canvas bags. The mailroom clerks had already abandoned the room, making their way up the narrow iron stairs to the lower deck.

Rollinson sloshed around in knee-deep water, doing the best he could to keep the huge furnace fired. It was self-preservation that kept him there when the others fled. He was no hero by nature and did not view staying behind as an act of courage. Rollinson knew that the longer he could keep the fires burning the more likely Blanchard would be able to keep the ship afloat. Besides, he also knew another way out of the boiler room that would take him up to the main deck. With the water swiftly rising Rollinson knew he had to shut the dampers and bank the fires. He tried to call up to the bridge but all communication on the ship was out.

The engine room crew scrambled up the escape ladders and laced their way topside. They made it as far as the cook's quarters where it was complete calamity. Pots, pans, dishes, glasses, giant cans of food, crates of vegetables, cartons of milk, everything imaginable that had been stored was bobbing around in a foot of freezing seawater. The head chef had long ago abandoned his quarters, moving his cooks and servers upstairs to the main. One of the cooks, his white T-shirt and pants soaked through, wandered aimlessly through the rubble. He was bleeding from the forehead where he hit his head trying to escape up one of the companionways.

"We're sinking," the disoriented cook cried.

Someone tried to help him, but the wild-eyed man ran down a staircase heading directly to the watery engine room. No one tried to save him.

> Huge waves, no doubt, tore off her superstructure and swamped the vessel, sending her and those aboard to the bottom of the sea. The terror experienced by those aboard can only be imagined. (Tom Seymour, *Fishermen's Voice*, October 2014)

29: E. Dudley Freeman

Former state senator E. Dudley Freeman stormed out of his stateroom and headed for the bridge. Gigantic waves washed over the foundering ship. Freeman had been in the Maine legislature in 1845 when it voted to incorporate the steamship company and always expected special consideration when it came to the company. When he reached the bridge, he demanded to know from Blanchard what was going on.

"Where are we? Why haven't we put into port?" Freeman insisted knowing. The stoic Blanchard advised Freeman to return to his stateroom and make sure his life preserver was secure.

Inches from Blanchard's face, Freeman screeched, "Are we going down?"

"Not if I can help it," Blanchard said.

He ordered one of the crew to escort him off the bridge. Freeman shook off his escort.

"You'll be hearing about this," Freeman bellowed as he stormed off the bridge back into the maelstrom. "This is the last ship you'll ever command. I'll see to it." he bellowed.

Blanchard knew there were forces far more powerful at play than E. Dudley, conspiring to make that happen.

Heading back to his stateroom Freeman ran into Eben Heuston. He demanded to know from Heuston if the life boats were going to be launched.

"Only Captain Blanchard can order life boats into the water," Heuston told him.

"Where are the life boats, boy?" Freeman demanded.

Heuston pointed aft.

"How many are there?"

"There are ten after on the port side," Heuston said. "But they are secured and only . . ."

Freeman wasn't listening. He clung to the railing and made his way aft. He tried as best he could to untie one of the life boats. A wave hit the ship knocking him headlong into one of the cast-iron pulleys that held the tiny life boats. Freeman lifted himself up off the deck, disoriented, and stumbled a few steps before another wave hit the boat shaking the vessel to its very core. Freeman lost his footing and was unable to grab anything to hold on to. In the cold cruel darkness Freeman was washed overboard, disappearing into the violently churning sea.

30: Plunged into Darkness

First Mate MacKey burst into the pilothouse. There were cracks in the bulkhead and the main beam had caved, he blurted out.

"The engine room is flooded," he said.

Blanchard kept his dark eyes on whatever faintly resembled the horizon.

"We'll stay the course," he said. "Get down below and get the bilge pumps working. Tell the crew to begin patching the leaks."

The first mate turned to leave.

"First Mate MacKey," Blanchard called after him.

"Yes, sir, Captain Blanchard," MacKey said, stopping in his tracks and standing at attention.

"Please button up your uniform. This is not an opportunity to break with protocol," Blanchard said. "Button your top button."

"Yes, sir," MacKey said.

He fixed the button on the collar of his uniform.

"As you were, First Mate MacKey," Blanchard said, letting the first mate go.

Hysterical passengers, bound up in their life jackets, men, women, and children, gathered bravely in their cabins. Some cried, while others tried to calm and console their young children. Others knelt and prayed. Some filed out onto the deck. Many hung on to ropes and rigging, anything that was secured to beams and pillars. Towering surf pounded the steamship, sending sheets of frigid foaming waves high onto the ship's decks. Fighting against the elements, they gathered in frightened mobs along the bow and stern of the ship. Inside the pilothouse the light flickered after another wave crashed against the ship broad side. There was another loud boom and another crashing wave. The lights on board the *Portland* flickered and went out. The whole ship inside and out was plunged into darkness. Hurricane lamps and candles were lit everywhere on board.

Most of the crew had found shelter below deck where they watched in horror as the waves washed over the decks in a relentless torrent. On the bridge, Blanchard's visibility was a white sheet of swirling sleet and snow. He could barely see ten yards ahead of him. An immense wave struck the ship, lifting it up as if it were a child's toy and spun it broadside. The railing along the aft gave way and crashed into the sea.

The crew frenetically tried to patch the leaks. To stem the tide, hot pitch was poured into cracks that were gushing water. As soon as one leak was patched another would appear. Icy water was flooding into the engine room and lower decks. The crew formed a line and tried to empty out the water using fire buckets. The last man in the line at the top of the stairs leading out of the boiler room emptied the bucket into the already flooded upper deck. The water poured down from the upper deck. Their best efforts were futile. The ship was going to pieces.

31: Inside the Saloon

The *Portland*'s saloon was all but empty except for a few hale and hardy fellows sitting around one of the last remaining tables left upright, drinking, enjoying a final cigar or cigarette, and playing cards.

"Drinks are on the house," the bartender said when the lights went out.

There was nothing any of them imagined they could do but ride out the storm. Most were old and single, and none had families, or at least none on board the ship. If they had, they would have, like the rest of the passengers, been huddled together with them somewhere. None of them wore life preservers either. They stayed bundled up in their coats and capes and top hats and derbies, their scarves pulled tight around their collars and their gloves on, even when dealing the cards. It was bitterly cold on deck and not much better inside, but at least they were protected from the wind and sheets of snow that pummeled the ship. No one dared speak of the ship or the storm or what might or might be coming next. They engaged in small talk about politics or sports. Why not ride it out in the comfort of good liquor and a fine cigar? After all, everything on board was free now. Cigars, cigarettes, and liquor were all at their disposal, or at least the bottles that hadn't tumbled off the shelves and shattered on the floor.

Several harried-looking ship's stewards and cabin boys were huddled together along the bar, engaging in the time-honored pastime of gossiping about the passengers. They didn't appear to be worried, or if they were, they didn't show it. Some of the group claimed they had ridden out worse storms at sea. Whether it was simply bravura or not didn't seem to matter. It was at least comforting to hear.

Another grinding jolt sent several of the passengers gathered around the upright table flying.

The men climbed off the floor, composed themselves, poured themselves another drink and continued on with their card gamc, by candlelight. Smoking, drinking, playing cards, but it was a tenuous and somber game.

Elsewhere on board, down the narrow corridors that led to the staterooms came the anxious murmurs of people, the sound of doors slamming shut, occasionally the rush of footsteps clamoring up or down the cast-iron staircases. Throngs of passengers, men, women and children, mingling together, made a strange picture. Their dress was an odd mixture of topcoats, top hats, derbies, bonnets, life preservers, evening clothes, fur coats, and turtle-neck sweaters.

Several passengers wore their life preserver over their bathrobe. Some clutched anxiously to a suitcase or bag of valuables. The anxious, noisy throng of passengers convened along the deck were in a jumbled disarray, overwhelmed by uncertainty and fear. The excitement that percolated through the crowd soon disappeared. The ship veered sharply to port. There was a breathless few moments of silence. The creaking woodwork, the rattle of glass breaking, the free fall of tumbling cargo and furniture suddenly stopped. Far more than the deafening jolt that had proceeded it, the sudden silence now stirred the passengers. The engines came to a complete stop.

32: First Mate John MacKey

First Mate John MacKey almost made it to the top of the escape ladder when he began to lose his grip. The icy water was rising fast behind him. His foot slipped on a rung. The ship lurched sideways. MacKey fell backward and found himself spinning round and round in a whirlpool of frigid water. He tried to cling to the railing but was being sucked down deeper. Kicking hard toward the surface, he tried to swim clear.

In the churning turbulence there was a web of thick ropes, errant deck chairs, wood planking, trunks, and every other sort of debris, swirling wildly around him. It was hard to see. Trying to reach the surface, his lungs nearly bursting, MacKey's head hit the corner of an iron winch, knocking him unconscious. His body slipped slowly down into the tempest. The cold, dark ocean poured in through the battered hull of the ship. He was gone.

33: Cold Comfort

Despite the pandemonium on board, Heuston did the best he could to offer comfort to passengers. Helping an elderly gentleman into his life jacket, Heuston stopped, knelt down on one knee and tied the old man's shoelaces.

He had to coax a hysterical young woman into her life jacket.

"I don't want to die!" the woman screamed.

"It's all right, Miss. You'll be all right," Heuston told her. "You must be brave. For the children's sake."

"It hurts," the poor woman sobbed. "I'm too young to die like this."

"And much too beautiful," Heuston said.

The poor woman's vanity superseded her fear.

"Do you really think so?" she suddenly asked.

"I've seen a great many beautiful women on this steamship and I can confess, you are among the most beautiful," Heuston said.

He made some adjustments to the belts on her life jacket.

"There," he said. "Fits like a glove."

The woman appeared placated for the time being.

Farther along the deck, he encountered a matronly woman. He helped her on with her life jacket.

"It's all the rage in Paris this year," he said, trying to console her. "And it will keep you warm."

For the most part, the passengers stood together in tight knots of family or friends, waiting or quietly pacing the deck. A few prayed along with Pastor Thomas Bryant, who was heading home to the seminary in Five Islands to continue his religious studies. The group huddled around Bryant prayed quietly. Others appeared lost in thought. Many held hands. The slant of the deck was so steep that people could no longer stand upright. Many held onto railings, or ropes, anything that could support them as the waves lashed against the stricken ship. Some cowered on the floor.

Heuston found Charlie Thomas, his wife, his daughter, and her dog standing just outside their stateroom. Gladys Thomas was holding onto her mother's skirt with one small hand and desperately clinging to the yapping dog with the other. The little girl was refusing to put on her life jacket. Heuston knelt beside the little girl.

"What's your doggie's name?" he asked.

"Pat," the little girl said, still clinging to her mother's dress while the puppy squirmed in her other arms.

Heuston laughed.

"Yes. Pat the dog. Well, I suppose we should put a life jacket on your dog as well," Heuston said.

He fastened a life jacket as best he could to the puppy, making the little girl giggle with delight. He easily convinced her to put one on as well.

Heuston moved on, but a giant wave crashed over the railing knocking him to the deck. He managed to get back up onto his feet and turned around to look, but the Thomas family were gone.

34: The Smokestack

Wave after wave pounded the ship, lifting it up as if it were a child's toy and spinning it sideways. The railing aft gave way and crashed into the sea. The crew frenetically tried to patch the leaks in the ship's hull. It was too late.

The tilt of the ship grew steeper, and suddenly slow, steady creaking noise was heard, growing louder and more resilient by the minute. Then a thunderous crack resounded. The ship's mighty smokestack had broken. Slowly creaking louder and louder, it began to bend, spewing sparks and hot ash. The fuming tower let out a final gasp and then fell, smashing through one of the side paddles and onto the starboard deck, splintering a huge portion of the upper deck and ripping through the ship. Sections of it burst into flames. Part of it struck the waves, sending a shower of sparks and releasing a billowing, hissing funnel of smoke shooting into the air.

The shock of what had happened sent the crowd into a state of panic and confusion. A roar of unbridled distress rose from the crowd. Ten passengers and crew standing along the starboard railing were lost when the smokestack crumbled. They were scattered into the sea. Others scrambled for safety. Those trying to catch a glimpse of what had happened and assess the damage were swept out to sea by a colossal wave that followed in the wake of the smokestack's destruction. The wave slapped hard over the top deck, sending even more passengers and crew to their doom.

Some came sputtering to the surface. Several of the victims were seen briefly, frantically trying to swim or stay afloat. The temperature of the water was about 30 degrees. In cold water like that, life jackets did no good. It was hard to see what was happening below. Some clung to railings and peered over the side to see what happened. There were bloodcurdling screams. Shouts. Pandemonium broke out. People were running for any cover they could find, wailing, some staring in disbelief, trying to take inventory of their loved ones and those snatched from the deck without warning. The *Portland* began listing heavily to port.

Shaken passengers and crew alike jammed together in a noisy, terrified horde, doomed prisoners of the storm. Many retreated up the deck ahead of the giant gaping hole where one of the ship's side paddles had been cut in two by the felled smokestack. The *Portland* swung around, controlled by the whim of the furious sea. The ship filled with loud popping, deafening thuds. A terrible thunder roared across the steamship as everything remotely moveable broke loose.

> Waves kept smashing the lower superstructure until there were only six to eight poles holding up the wheelhouse in front of the ship. The whole deck of the ship got ripped off. (Art Milmore, *And the Sea Shall Not Have Them All*, 2017)

35: Rollinson's Escape

Harry Rollinson bolted down the gangway just ahead of a swarm of passengers who were running the same way, trying to clamber up onto the top deck and bridge where the water hadn't yet reached. He kept ahead of the crush and vaulted up the steps two at a time to the top deck. The *Portland* twisted violently portside. The sickening turn threw most of the people behind him into a huge heap along the gangway. The top deck was listing too steeply to stand on. Rollinson slid to the starboard rail. He worked his way up the side, still holding onto the rail until he reached what appeared to be safety. Throngs of helpless people were gathered there.

He had barely escaped the flooded engine room. The pumps couldn't keep up with the gush of water pouring into the lower decks. The sea was pouring in so fast that the air was being forced out of the lower compartments producing tremendous pressure. The ventilators rained down on him as he tried to escape.

Before he could free himself, he was sucked under, spinning round and round in a whirlpool of freezing water. He was pulled back into the rushing tide and pinned to a wire grate of an air shaft. He prayed the grate would give way before he was sucked down into the tumult below him. Using all his strength, he pulled the grate off and crawled into the air shaft. Gasping for breath, his strength all but spent, he managed to claw his way up the shaft onto the lower deck.

Sprinting up the gangway ahead of the crowd, he made it safely to the upper deck just as the huge smokestack gave way and crashed into the sea. He was thrown clear of the cataclysm, lucky to be alive.

> Between 11 and 11:45 p.m. *Portland* was sighted three times, but this time to the southeast of Thacher's Island–she was being driven south by the storm. When sighted at 11:45 p.m., she is said to have shown severe storm damage, especially to the superstructure. By this time conditions on the steamer must have been dreadful, and all aboard must have known they were in grave danger. (Judi Heit, "The Gale of 1898," *The Portland Gale*, http://portlandgale.blogspot.com/, 2010)

36: Chaos

> And now there came both mist and snow,
> And it grew wonderous cold:
> And ice, mast high, came floating by,
> As green as emerald.
>
> (Samuel Taylor Coleridge,
> "The Rime of the Ancient Mariner," 1798)

In the chaos nobody knew what happened. Some had been swept overboard. Those that were left huddled together like swarms of angry, confused bees racing to the top deck. Everything was cast in a murky iridescent glow.

Second Mate Frank Patterson burst onto the bridge.

"She's taking water!" he blurted out.

Captain Blanchard was unmoved.

"Where is First Mate MacKey?" he asked. "I won't tolerate dereliction of duty."

"He's gone," Patterson said.

"Well, find him," Blanchard ordered.

"I think he's dead, sir. Washed overboard," Patterson informed him.

"Second Mate Patterson, you are officially promoted to First Mate," Blanchard said.

Patterson was unmoved by the sudden promotion.

"Sir, the ship is going down," he told Blanchard. "What should we do?"

Blanchard didn't flinch.

"She's not going down," Blanchard said confidently. "We will ride out the storm in safe waters."

"But sir," Patterson said. "Three of the four decks are flooded. The engine room is gone. The smokestack's fallen. We are taking on water everywhere."

With the side paddle damaged and half the steam, Blanchard did his best to keep the ship righted. The extra bilge pumps below had begun to slowly lower the water level, but even with them it wasn't enough. Water was seeping up through the metal floor plates and gushing in from the splintered seams.

"Fire the flares," Blanchard commanded. "Fire them at intervals of every ten minutes."

"Sir, there's no one there that will see us."

Blanchard didn't answer.

He took his clean handkerchief from his pocket and wiped the brass railing along the wheel housing.

"First Mate Patterson, you have your orders," he said.

It was after midnight, approximately 12:30, when the first flare shot up into the dark sky from the bow of the stricken *Portland*. The flare burst into a red flame followed by a shower of glittering sparks falling into the sea. The passengers looked to the heavens as the arc of the flare soared into the dark sky and exploded. Some were tempted to ooh and ahh at the display, as if they were watching fireworks.

37: Man Overboard

Harry Rollinson made it to the saloon. Barely a handful of men were still there. Charlie Johnson, the saloon keeper was still behind the bar when Rollinson came racing in. He was covered from head to foot in thick black coal soot, caked onto his body from the sea that had flooded into the boiler room.

Rollinson wasted no words. He marched into the saloon, strode behind the bar, grabbed a bottle of good whiskey from the top shelf and a handful of cigars.

Charlie Johnson tried to stop him, but he was no match for the hulking Rollinson.

"I will call the captain," Johnson threatened. "This is for passengers. Not crew."

"Call captain," Rollinson grinned, as he lit up his cigar. He opened the bottle of whiskey and slugged down a huge mouthful.

"You'll be reported when we reach the mainland," Johnson said. "You won't ever work again."

No one else in the saloon said a word. They had other concerns.

Rollinson glared at him.

"We'll never see land again," Rollinson bellowed. "You can't be as stupid as you look." Rollinson took another long drink from the bottle. The ash on the end of his cigar glowed bright red. "Look around." With that, Rollinson took the bottle and his cigar and bolted out the door of the saloon.

The ship rocked violently. Along the deck, Eben Heuston hung on for dear life. He trusted Captain Blanchard. He knew if anyone could ride out a storm like this, it was Blanchard. Still, being a religious man, he prayed quietly to himself, believing that God could do what Blanchard might not–save them all.

Heuston thought of his wife and his young stepchildren. He thought of their marriage, now exactly one year old and of how much it would hurt Margaret if she were to lose another husband to the angry sea. He prayed she wouldn't. He opened the small jewelry box he had tucked in his vest pocket and looked admiringly at the small gold cross he intended to give to her as an anniversary present.

The roar of thunder was undistinguishable from that of the waves that hit relentlessly against the ship. Up ahead of him, clinging to an outside railing, Heuston saw a familiar face, Harry Rollinson. Rollinson had stripped down to his underwear and was puffing on a long cigar. In one hand he held a bottle of whiskey, in the other he had a gob of black engine grease. At his bare feet was a cannister of the grease. He was furiously smearing the grease all over his body. Rollinson was lubricated inside and out–with whiskey and engine grease to keep his body from freezing.

"What are you doing?" Heuston asked as the ship spun wildly.

Rollinson pointed to what looked to be a flickering light off in the distance, barely visible in the torrent of sleet and snow.

"I'm swimming to that. It's got to be land," Rollinson said, the tip of his cigar glowing bright red.

"It could be anything. A mirage," Heuston cautioned him.

"Ain't no mirage. It's land. I'm taking my chances," he said.

Rollinson heaved his massive frame up and over the side of the railing and hung on, facing Heuston.

Somehow forgetting the severity of the situation, Heuston said, "That cigar won't stay lit."

"I get it lit on the other side," Rollinson said. "Tell my mother I love her, and tell Blanchard . . ." his voice trailed off as he looked below. "Tell him to go to hell."

Rollinson let go of the railing and fell feet first into the crushing waves below, flapping his arms as if somehow they would carry him away off into the stormy sky like a bird. But they didn't. He fell into the pounding surf and disappeared beneath waves.

38: The Calm before the Storm

The crippled ship, like a decaying wooden skeleton, was tossed helplessly on the sea. It was just past 1 a.m. Sunday morning and the *Portland* had been driven south to the tip of Cape Cod, some ten or twelve miles off the coast of Wellfleet. There was barely anything left of it. Its enormous side paddle was demolished. Its smokestack had collapsed. The bridge and some portions of the upper deck were all that was left above water. The ship listed to port. The sea was washing onto the bridge.

The *Portland*'s bow slipped deeper into the stormy water and her stern swung up into the bitterly cold night air. The ocean in every direction was littered with boxes and cans, deck chairs, crates, planking, life jackets, and debris that bobbed to the surface. Hundreds of passengers had been swept off the ship and thrashed at the water, clinging to whatever wreckage they could find or to each other. Their cries died away quickly in the freezing water. Then there was no trace of life in the sea.

The remaining crew and passengers had gathered on the top deck trying desperately to find shelter. Eben Heuston had made it there, as had George Kenniston Jr. and Emily Cobb. Kenniston had tied a thick curtain rope that had hung in his stateroom around his waist and tied the other around Cobb's, so they wouldn't get separated in the storm. She still grasped her satchel of sheet music and he had his leather-bound journal tucked into his belt.

Blanchard spoke to the few members of his crew gathered along the bridge.

"Well, boys, you've done your duty. No one can fault you. Look out for the men and women. Then, I guess it's every man for himself," Blanchard said. Sea foam swirled over the top of his shoes.

"What's done is done," he said peering into the darkness. "Save yourselves."

The night had passed slowly. When morning came, miraculously, the *Portland* was still afloat.

According to Cape Cod Life-Saving District Superintendent Benjamin C. Sparrow, "Between 9–10:30 a.m. there was a partial breakup of the gale. The wind became moderate. The sun shone for a short time and the atmosphere cleared."

Blanchard ordered the crew to fire flares. They shot up into the sky from the bow of the stricken ship. A thick impenetrable fog engulfed everything. The flares soared into the dark sky and exploded. The storm had moved inland, striking up and down the New England coast where it tore across the region for another full day. The sea was still choppy. Bitter cold waves splashed against the wreckage.

Captain Samuel O. Fisher of the Race Point Life-Saving Station in Provincetown, Massachusetts, reported that at 5:45 a.m. he heard four short whistle blasts from a steamer. Without any visual identification there was no way of knowing if the blast came from the *Portland* or some other distressed ship off the coast.

39: Full Fathom Five

Full fathom five thy father lies . . .

—William Shakespeare, *The Tempest*

Somewhere out on the dark sea, shrouded in the fog, was the schooner *Addie E. Snow.* Despite its heavy cargo of granite, it had been tossed and turned, pitching wildly in the angry waves and was now rudderless. Like the *Portland*, it too had been blown far off course. It was blown southward in the gale toward the "bared and bended arm" of Cape Cod.

Long considered the "graveyard of the Atlantic," because of its treacherous shoals and currents and the number of ships that had wrecked there, Cape Cod was the last place on earth the captain of the heavy schooner wanted to be. In fair weather it was dangerous. In a storm of this magnitude it was deadly. Like the *Portland*, the *Addie E. Snow* was running without lights or a means of communication.

The captain of the *Addie E. Snow* had taken to sounding the ship's bells at three-minute intervals and setting off flares, but in the turbulent storm it was useless. Still, he had his crew fire off flare after flare and ring the ship's bells aft and starboard. His crew had been on lookout all night and never saw anything, not even the distressed and battered *Portland* that came within a hundred yards of it. Blanchard's crew, too busy trying to keep the ship afloat, never saw the fiery red flares of the *Addie E. Snow* careening toward it in the fog along the outer banks of Cape Cod.

Eben Heuston was glad to be alive. He took the gold cross he had bought for his wife for their wedding anniversary from his pocket. He kissed the cross and placed the gold chain that held it around his neck for safe-keeping.

George Kenniston Jr. and Emily Cobb were still bound together around their waists by the thick curtain rope. She was shaking. Kenniston did his best to comfort her. They were both soaked to the bone. Blanchard peered out into the fog and saw nothing. He had managed to keep them afloat through the storm. He held out hope that perhaps they could be rescued.

There was no warning, no sound. The rudderless *Addie E. Snow*, the huge schooner, weighted down with granite, barreled out of the fog, slamming into the midship of the *Portland*, tearing through what was left of the ship's bridge and breaking the ship in two.

Both ships sank in a matter of minutes. Passengers and crew were swept into the sea. Their cries soon died away. Then there was no trace of them. Everything became strangely silent. It was over.

> Very few mariners at the time considered the possibility of a collision between the *Portland* and the 162 ton schooner *Addie E. Snow*, but later the brother of the schooner's captain examined some of the *Portland*'s wreckage on the beach. To his surprise, Captain Brown's brother found the *Addie E. Snow*'s medicine chest in the *Portland* litter and afterwards found several other articles which had been on the schooner. (Edward Rowe Snow, Jeremy D'Entremont, and William Quinn, *Storms and Shipwrecks of New England*, 2005)

Seafaring historian and author Edward Rowe Snow, who wrote extensively about New England storms and shipwrecks, called the sinking of the *Portland*, "the worst maritime tragedy of the 19th Century in New England waters." And, according to Dr. Craig MacDonald, NOAA's Stellwagen Bank National Marine Sanctuary Superintendent, "The wreck of the *Portland* has never lost its mystique. It's New England's *Titanic*."

The small schooner *Addie E. Snow* was also lost during the storm, and her remains lie less than 1/4 mile from *Portland*'s grave. It is thought that the two vessels may have collided, hastening their ends. (Andrew Toppan, *Haze Gray & Underway*)

The brutal winter storm had pounded the New England coast for two days, from Saturday, November 26, through Sunday, November 27, piling up twenty feet of snow in some places, shutting down train and shipping lines, and knocking out all communication from Cape Cod to Maine. By Monday, November 28, it was over, but the horror of what it had left in its wake was just beginning to surface—entire fishing fleets sunk, homes and businesses demolished, piers washed into the sea, entire coastline villages swept away. But that was only the beginning. For the families of those on board the *Portland*, the worst was yet to come.

BOOK THREE

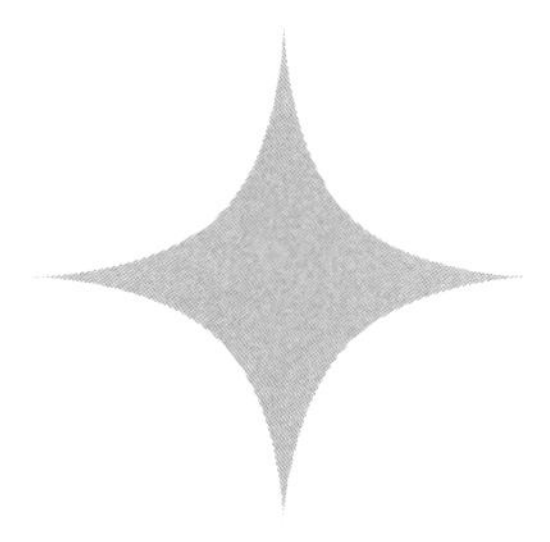

The Wreck

"The Steamer *Portland* Surely Lost"

–*The Boston Globe*, November 29, 1898

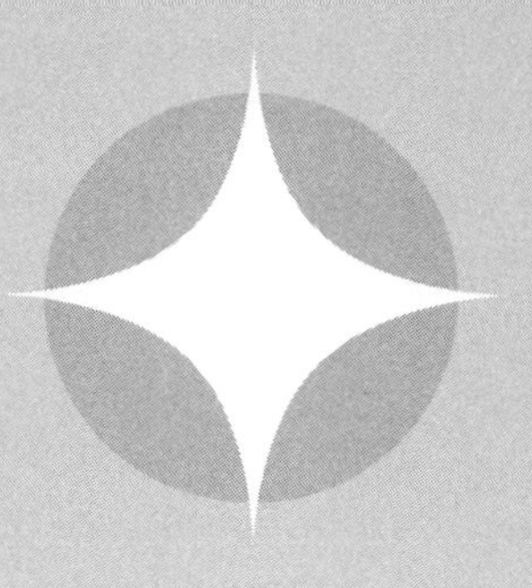

1: All Lost

"They're all lost," Jack Johnson told the first newspaper reporter who interviewed him. It was Sunday night, November 28. Frank Stanyan was from the *Boston Globe*. Johnson was a Cape Cod Life-Saving Service "surfman" assigned to the Peaked Hills Bar Station in Provincetown, Massachusetts. Johnson was a wizened old Cape-Codder with a crumbled, weathered face who had, over the course of the many long years tending lighthouses and volunteering as a lifesaver, taken on the appearance of a stealthy old crab, even to his sideways gait when he walked.

The Life-Saving Service was made up of volunteer "surfmen" who occupied various lifesaving stations along the Cape Cod coast. They kept watch for ships in distress and leaped into action when necessary, using surfboats, buoys, and rescue lines to help rescue wreck victims. The Peaked Hill Bar Life-Saving Station was a large boathouse and dormitory, built in 1872. The Cape Cod Life-Saving Service, established in the 1870s, later became the United States Coast Guard.

At about 7:20 p.m., shortly after the storm had finally subsided, Johnson was patrolling a section of the windswept Truro beach near the very tip of Cape Cod when he made a discovery. As he scanned the beach with another surfman, he saw something being tossed up onto shore in the incoming tide. With only a lantern to guide him, Johnson fought his way into the surf, reached down and snatched his find from the incoming wave. In the dim light of his lantern he and his partner examined the object he had wrenched out of the sea. It was a cork-filled life preserver with the words, "Steamer Portland of Portland," printed on it in big red lettering, the number 40 was emblazoned below it. A short time later, he found a can stamped with the words "Turner Center Creamery of Maine." Other similar cans followed. They were the same industrial-size kind used aboard the *Portland.*

Lifesavers like Johnson spent endless hours patrolling the shore during and after storms and at night, when wrecks were more likely to occur. Johnson had been around the sea long enough to read its signs. He knew that the sea ultimately gave up its secrets, even the most horrible of them. He braced for the worse.

> I was bound west toward the station when I found the first thing that landed from the steamer. It was a lifebelt and it was one-half mile east of the station. At 7:45 o'clock that evening I found the next seen wreckage, a creamery can, 40-quart, I guess. It was right below our station and nine or ten more of them, all empty and stoppered tightly came on there closely together. (Jack Johnson, November 28, 1898)

"What's the forty mean?" Stanyan, the newspaper reporter, asked when Johnson showed him the life preserver.

"The number of the jacket. All the jackets are numbered," Johnson told him.

Johnson's heart grew heavy with grief as he showed Stanyan the life preserver and the creamery cans he had gathered along the shore. He knew full well what had happened, and it shook him to his core, this despite his many years of hard service at sea, or perhaps because of it. There was no way to confirm anything with the lines down and all other forms of reliable transportation at a standstill.

The men examined the debris and anticipated that there would be much more to come. They would not be wrong. Johnson assured the young reporter that there would be more wreckage soon, and bodies. He knew that the question would become, "How many bodies?"

The lifesaving stations were manned by the most expert surfmen and boathandlers to patrol the coast at night and during storms and bad weather. The men of Provincetown, Massachusetts, were recruited for their demonstrated ability in boat handling. The stations were manned ten months a year, from the first of August until the following June. The keeper of the station was on duty throughout the year. The stations were plain structures designed to serve as a home for the crew and to provide storage for the boats and equipment. They were set back as far as possible from the ocean and painted red so that they might be identified from a long distance at sea. There was a lookout or observatory where the surfmen would keep an account of shipping traffic during the day and a sixty-foot flagstaff used to signal passing ships using the international code. There were nine lifesaving stations on the Cape, including Race Point, Peaked Hill Bar, Cahoons Hollow, Orleans, Chatham, Highlands, Pamet, Nauset, and Monomoy Point. Just one of these original lifesaving stations, the Cahoons Hollow Life-Saving Station, built in 1853 and reconstructed in 1897 after a fire burned it down, remains at its original location. The stations were decommissioned in the 1940s. Those stations manned by Provincetown men and the members of King Hiram's Lodge were Race Point and Peaked Hill Bar, Highlands at North Truro, Pamet at Truro, and Cahoons Hollow at Wellfleet.

The Peaked Hill Bars Life-Saving station, where Jack Johnson was assigned, was located two and a half miles east of Provincetown. It was considered one of the most dangerous stretches of Atlantic coastline. According to James J. Theriault, curator of King Hiram's Museum in Provincetown, "Hundreds of ships have been lost there, spilling men's lives and ships' cargoes onto its shores."

It was not until approximately 11 p.m. Sunday night that the deluge of wreckage from the *Portland* washed ashore. Broken, battered pieces of debris from the ship; mattresses, chairs, windows, doors, paneling, even lightbulbs, all began to wash ashore everywhere along the Cape Cod coastline from Wellfleet to Provincetown. Lifesavers along the Cape Cod coast began collecting what they could salvage from the wreck.

> According to a Boston newspaper account on the day following the storm, "Miles of coastline from Buzzards Bay to Cape Ann are strewn with wreckage. A blanket of numbed sadness smothers coastal New England this morning." Only 36 bodies from the *Portland* were ever recovered on nearby beaches. Many claim wind-up wristwatches on some of the victims were stopped at 9:15. If true, was it 9:15 a.m. or p.m.?" (Louis Varricchio, "The Wreck of the S. S. *Portland*: The *Titanic* of New England," *Sun Community News*, 2012)

And, just as Johnson predicted, the sea soon gave up its dead. Fifteen bodies washed ashore Monday morning. Ultimately, thirty-six bodies in all would be pulled from the waves.

> As she took her final plunge the superstructure was probably torn away from the main deck and was smashed to kindling wood. Those inside were thrown into the icy water as the wooden deckhouses disintegrated, some being killed outright by falling beams and other debris, others being caught in the wreckage and carried under the surface to drown. (Edward Rowe Snow, Jeremy D'Entremont, and William Quinn, *Storms and Shipwrecks of New England*, 2005)

2: Always Ready

Frank Stanyan from the *Boston Globe*, like every other newspaper reporter who converged on Cape Cod to cover the tragic story of the *Portland*, depended on the lifesaving crews for information. The Cape Cod lifesaving crews were working day and night recovering bodies ever since Jack Johnson first pulled the *Portland* life jacket out of the surf.

The official maxim of the United States Coast Guard is Semper Paratus, Latin for "Always Ready." However, the Guard's unofficial motto has always been: "The book says you have to go out. It don't say you got to come back." On the night of the great New England storm, the Guard and its lifesaving units along the dangerous Cape Cod coast were kept frantically and dangerously engaged.

Cape Cod was infamously known to all seamen as "the graveyard of the Atlantic." None of the sailing charts for Cape Cod were trustworthy because of the shifting sand bars. Charts had to be updated yearly in order to stay accurate, but they weren't. Mariners choosing to navigate around the Cape took their lives and those of their passengers in their own hands in fair or foul weather.

> Since men have taken to the sea in New England, Cape Cod and its shoals have been known as a menace to mariners. The Cape has well earned its reputation as a graveyard of the Atlantic Ocean. . . . According to Cape maritime historian and shipwreck author William Quinn, [it] was once called "Dangerfield" because of the treacherous waters off its coast. Left by receding glaciers thousands of years ago, the Cape sticks out into the path of ships plying waters to and from the major ports of the northeast, including Portland, Boston, Providence and New York.
>
> For centuries, captains have tried to steer well clear of the grasping sandbars and shoals that have snared thousands of ships and claimed the lives of hundreds. . . . No ship nor sailor, no matter how stout or how knowledgeable, was immune. ("Cape Cod's Reputation as a Graveyard," *Cape Cod Times*, 2011)

There are many records connected to the *Portland* in the logs of the Cape's nine Life-Saving Service stations. According to one set of records, "At the Cahoons's Hollow station, on November 28, 1898, the body of George Graham, a Negro, washed ashore. He is identified by name in the log entry, and he is the only crew member whose body was identified at the time of recovery."

Joshua James, of the Point Allerton Life-Saving Station in Hull, Massachusetts, was one of the Life-Saving Service's most decorated and heroic members. He began his lifesaving career n 1842, when he joined the Massachusetts Humane Society and went on to become famous as the commander of civilian life-saving crews in the nineteenth century. James was so instrumental in a plethora of rescues over the years that he was awarded a special silver medal by the Humane Society in 1886 that cited him "for brave and faithful service for more than 40 years. . . . During this time, he assisted in saving over 100 lives. In the great storm of 1888, he and his men saved twenty-nine people from six vessels." For his heroics, he received the Gold Life-Saving Medal from the U.S. Government. During the gale of 1898 he rescued two survivors from two vessels that wrecked on Toddy Rocks in Hull; seven men using a breeches buoy from a wrecked schooner; five men from a barge beached along the coast, and three men from Black Rock Beach, a stretch of shoreline around the outer side of South Hull. During the thirteen years he was keeper of the Point Allerton Station, he and his crew saved 540 lives and $1,203,435 worth of estimated value of ships and cargo. James died at the age of seventy-five while on duty. He was buried with a lifeboat for a coffin. His tombstone bears the Massachusetts Humane Society seal with the inscription, "Greater love hath no man than this–that a man lay down his life for his friends."

> Here and there may be found men in all walks of life who neither wonder or care how much or how little the world thinks of them. They pursue life's pathway, doing their appointed tasks without ostentation, loving their work for the work's sake, content to live and do in the present rather than look for the uncertain rewards of the future. To them notoriety, distinction, or even fame, acts neither as a spur nor a check to endeavor, yet they are really among the foremost of those who do the world's work. Joshua James was one of these. (Sumner Kimball, Superintendent of the U.S. Life-Saving Service, 1902)

Lifesavers like James and Jack Johnson spent an endless amount of time patrolling the shore during storms and at night, when wrecks were more likely. The shore patrols were strenuous and exhausting. Distress calls from stranded ships propelled them into immediate action. Lifesaving crews would haul the lifesaving equipment by hand or by wagon. It was tough going in the sand and storms. From the shore they were prepared to launch their rescue boats, which were small but sturdy vessels, called surfboats. When the Life-Saving Service was first created the government supplied them with wooden flat-bottomed boats, but the boats proved useless for launching into the pounding Cape surf during a storm. The Service soon exchanged these for boats built by Cape-Codders who knew the waters. The new Cape Cod rescue boats were smaller than a whale boat, but similar in style, equipped with a keel, and narrowed at both ends and deep. More important, the new rescue boats were lightweight and could be launched in the Cape Cod surf and breakers. The lightweight rescue boats accommodated five lifesavers at the oars and a helmsman. The rescue boats were only big enough to save five people at a time, which required the lifesavers to make a multitude of trips in the treacherous storm-tossed waters to rescue stranded crews and passengers.

3: Fate Unknown

On Sunday morning, the whereabouts of the *Portland* was still unknown. Although wreckage and bodies from the ship began washing ashore late Sunday night, there was no way for newspaper reporters to report back the news because the storm had wiped out all means of communication–telephone and telegraph–from Cape Cod. The storm-battered coast of Cape Cod was in complete isolation.

At the steamship offices in Portland, Maine, general manager John Liscomb tried to remain calm. Liscomb was under the assumption that his telephone call to Boston earlier on Friday, ordering Captain Blanchard to stay in port until nine o'clock that evening had been followed. When Liscomb called Boston later that evening and discovered that the steamship had sailed out of port at seven o'clock as scheduled, he began making a series of telephone calls to ports along the route from Boston to Maine searching for the vessel.

Liscomb and others remained hopeful that when Blanchard encountered the first inkling of the storm, he put into port somewhere along the way. By the time Liscomb began making his telephone calls late Saturday night, many of the telephone lines had already been blown down. It wasn't until early on Sunday morning that he was able to make telephone contact with Gloucester, Massachusetts, a potential safe haven for Blanchard along his route back home, but the *Portland* had not arrived there. Still, Liscomb held out hope that the steamship had made it into some safe port along the way and that when he could, Blanchard would be reporting in that everyone was safe and sound. The call from Blanchard never came.

When the *Portland* failed to arrive back home, and no word of it came from any port along its route, a deadly silence fell over the Maine seafaring city. Liscomb issued a statement on behalf of the steamship company that appeared in the Portland and Boston newspapers:

> People should not become alarmed for the safety of the vessel as yet. Telephone and telegraphic communications with ports along the north shore is entirely cut off, and for that reason no word can be obtained as to the *Portland*'s whereabouts. It is not unlikely that the big steamer headed back for Boston when she ran into the storm Saturday night and is now anchored in a sheltered spot down the harbor, for no tugs dared ventured very far down on Sunday, and, owning to falling snow, a good view of what was down below could not be obtained.

The statement did nothing to quell the fears of families and friends who converged on the steamship company's offices in Portland and Boston in droves, frantically looking for some word concerning the whereabouts of their loved ones. No one could allay their mounting fears.

4: The Boston Ticket Office

Hundreds of frantic, worried family and friends converged on the steamship offices on India Wharf in Boston trying to find out the fate of their loved ones. Men, women, and children swarmed down to the tiny ticket office to find out exactly who was, or was not, on board the *Portland* when it sailed on Saturday just prior to the storm. Some tearful and most anxious, they wanted only to know if there had been any word of the missing vessel. Others asked to know if their loved ones had been on board or whether they had decided to cancel their trip back to Maine.

Still others came to the ticket window with fury in their hearts, pounding their fists on the window and demanding to know why the *Portland* had even left Boston Harbor in the face of the coming storm. There was only one employee in the steamship's Boston office, a clerk, William Barlow, and he had no answers for any of them. The harried clerk, his shirt sleeves rolled up, his eyeglasses perched on the top of his bald head, tried as well as he could to help everyone who came to the window, but it was an impossible task. There were hundreds of questions, and no answers to give.

In Boston the damage had been severe, but the city dug out quickly. The streets were practically cleared or at least made passable by the huge horse-drawn city snow plows by Sunday afternoon. Stores and businesses were open, and life in the city had once again resumed its hustle and bustle. But everyone seemed to be searching for some bit of news about the missing *Portland*. In both Boston and Portland people waited and prayed and hoped for some word of their missing loved ones.

Barlow apologetically informed people that there was still no word on the *Portland* and that it could be days before a complete list of the passengers on board would be known, since the only manifest of passengers was on board the *Portland*. There was no way of knowing who had left or who had decided to stay in Boston and ride out the storm.

"It's one of the safest ships we have," Barlow tried to console worried family members. "Captain Blanchard is one of our most knowledgeable and competent captains."

Barlow's assurances fell on deaf ears.

5: Making a List, Checking It Twice

William Barlow was at wit's end when part-time *Boston Journal* reporter Frank Sibley showed up. Telephone communication within Boston had been restored, and besides the hordes of people at the ticket window clamoring for information, the phone in the small office was ringing off the hook.

Sibley was assigned the impossible task of trying to piece together the *Portland*'s passenger list for the newspaper. Since the only manifest of passengers was on board the *Portland*, it was an impossible assignment, but Sibley was smart enough to realize that even though he couldn't get a passenger list from the clerk, he could piece together an unconfirmed one based on the hundreds of inquiries that were flooding into the small office.

If someone was calling or asking about a relative or loved one, Sibley could be fairly sure that the person being sought had to have been on their way to Portland that fateful night. With the names provided from the inquiries, he could begin to piece together a hodge-podge of names. It wasn't much, but it was at least something.

Sibley offered to help the beleaguered Barlow by answering the phone. Barlow welcomed the help. Sibley took up a position inside the office beside the telephone. With his notebook in hand, Sibley answered the telephone and vigilantly jotted down the names and addresses of the callers, as well as the names of the people they were looking for. Little by little, he was able to cobble together a list of names. It was like putting together a jigsaw puzzle in the dark, but that's all he had to go on. Piece by piece, Sibley's ingenious plan produced the first and only list of passengers thought to be on board the still-unaccounted-for *Portland*. Although the accuracy of the list remained questionable, it was at least something, and as time wore on, anything was better than nothing. That night, the *Boston Journal* ran a special edition with Sibley's story on the front page. It was the only newspaper in the city to publish a tentative list of passengers on board the missing *Portland*.

ONLY PORTLAND PASSENGER LIST ONBOARD VESSEL

Number on Board Still Unknown

BOSTON–The exact number of persons who were carried away from Boston by the *Portland* will probably never be known, as no list of passengers was retained on shore when the vessel left last Saturday. Many estimates of the number onboard have been made, but the estimates seldom agree.

Portland Steamship Company authorities place the total number of persons on the steamer at 100 or possibly 105. This estimate, however, is generally regarded as rather small. It has been stated that the number was as high as 155. It is more probable that 120, including passengers and crew, is near the correct number. The following passenger list was compiled as accurately as the circumstance permitted, but it is possible that some of those named were not on board the vessel. (*The Boston Journal*, November 28, 1898)

After Sibley's story ran in the *Journal*, the *New York Times* released its own ominous story about the possible fate of the *Portland*:

BOSTON. Nov. 28.–The managers of the Boston and Portland Steamship Company, stated to-night that there were grave doubts as to the safety of the steamer *Portland*, which sailed from here Saturday night.

Every harbor between here and Portland on the north shore has been heard from, and one of the south shore, and not in any case has the steamer been seen. The only remaining harbor which she could have reached is Provincetown, on Cape Cod, and news from that port is anxiously awaited. It is still impossible to reach Provincetown by wire.

The *Portland* carried sixty-five passengers and a crew of fifteen men. The passenger list is aboard the *Portland*, and at present

> there is no means of knowing the names of those on board. . . . Arrangements have been made with the Government to dispatch the revenue cutters *Dallas* from this port and *Woodbury* from Portland to hunt up the steamer.
>
> The *Dallas* will cruise along the south shore as far as Provincetown, while the *Woodbury* will make a long circuit from Portland around Cape Ann, and if neither of these steamers succeed in locating the missing vessel there is little hope of her ever being seen again.
>
> The *Portland* is a side-wheel wooden vessel of 2,284 tons gross. She is 280 feet long, 42 feet beam, and 15 feet depth. She was built at Bath in 1890. (*New York Times*, November 29, 1898)

6: Waiting Is the Hardest Part

Back in Portland, high atop Munjoy Hill, Margaret Heuston and others whose husbands or wives had been part of the *Portland* crew gathered at the Abyssinian Church for an all-night vigil, where they waited and prayed for the safe return of their loved ones. There were nineteen men and women from the church on board the missing vessel.

Judith Cobb, the widowed mother of Emily Cobb, waited in Portland for some word of her daughter. Emily was scheduled to make her solo singing debut at the Unitarian First Parish Church on Congress Street that Sunday, and she would not have missed it for anything in the world. Judith Cobb was certain that her daughter was on board the missing steamship. The Reverend John Carroll Perkins, associate pastor at the church, stayed with Judith Cobb, trying to comfort her during the long ordeal.

Boothbay banker George Kenniston Sr. waited with his wife in Portland for some word of his son, George Jr., who was returning from Boston after visiting his sister for the Thanksgiving weekend.

Captain Blanchard's wife, Emma, drove alone by horse and buggy from their home in Deering, Maine, to the steamship office in Portland to await news of her missing husband. She left their two children in the care of relatives. She was sure that given her husband's proclivity toward caution and safety, if there truly had been any possibility of danger, he wouldn't have left port. She knew this with all her heart and soul. And even if he had set out ahead of the storm, he would have put into the nearest port along the way if the ship and passengers were in any kind of danger. She waited in the steamship offices along Portland's Franklin Wharf for word.

Hundreds of others, just like them, prayed and hoped and waited for some word of their missing loved ones. But there was no word. Not for three, long, angst-ridden days—only dead silence.

7: Swept Off the Face of the Earth

> Black are the brooding clouds and troubled the deep waters, when the Sea of Thought, first heaving from a calm, gives up its Dead.
>
> —Charles Dickens, 1842

It was as though everyone on Cape Cod had been swept off the face of the earth, leaving the otherwise picturesque peninsula as desolate and barren as the face of the moon. Not a soul was in sight. Not a sound. Not even the once-roiling sea made a noise, as if out of respect, as if it had grown discomfited at its own uncontrolled rage. As if even it knew it had gone too far. It was the kind of dead silence that often follows any storm, when the worst is over.

For two days and nights before, the rocky seas had erupted with towering waves and pounding surf all along the Cape Cod seashore, smashing boats, washing away roads and rails, parts of homes, stores, and piers. From the tip of Cape Cod at Provincetown, Massachusetts, to the farthest points along the rocky Maine coast, the storm had wreaked havoc of epic proportions. At first, once the storm had ended, no one really knew the extent of the destruction. There was no way for the outlying areas to report back to places like Boston,

Portland, or Providence. Word of the cataclysmic destruction spread first only by word of mouth from one neighbor to another, from one small town to the next, none of it reaching the state capitals where news could be spread out among the populace in the various large daily newspapers or telegraphed to Washington, D.C.

Frank Stanyan, a reporter for the *Boston Globe*, went down to the hard, frozen tundra that was Cape Cod after the storm. In Orleans, an eerie, dead silence engulfed him as he trudged knee-deep through the snow and debris, his shoes sinking into the mounds along the narrow main road. Nothing was the same as it had been. There were fallen trees limbs strewn across the road, sunk deep in the snow, some resting atop the roofs of small battered homes and whitewashed stores that lined the main street. Telegraph lines felled from the storm lay in snake-like coils in and out of the snow mounds, some still lashed to telegraph poles, some poles snapped in two. In some places, the snow was stacked up higher than the buildings it rested against, concealing whatever store or business it might be.

Unrecognizable bits and pieces of shattered homes, shreds of lumber, broken glass were scattered everywhere for as far as the eye could see. And there was the cold–a bone-chilling cold made visible in the great billowy gusts of breath that heaved from Stanyan's nostrils and mouth as he trudged down the desolated and desecrated main street in Orleans. There wasn't a soul in sight. They were all gone, out searching for wreckage and, now, bodies. The idea of it made him shudder even more than the cold did.

Stanyan had taken the train from Boston down as far as it could go, to Sandwich. Rail service was washed out everywhere. He then took a carriage as far as the roads would let the carriage pass. He took a sleigh into Orleans. Cape Cod was in chaos. It was as if the great sea god, Neptune had risen up from the depths, stuck his mighty trident into the Cape and shook it, like a farmer might shake a fork-full of hay, and then toss it over his shoulder sending the fragile strands flying in all directions.

8: Trans-Atlantic Relay

Stanyan had arrived in Orleans on Sunday night, November 28. Now it was nearly three days since the *Portland* had sailed from Boston. He met first with Jack Johnson, who showed him the life preserver from the *Portland* and the creamery cans. Then word came that bodies were being washed up on shore. A base camp for newspaper reporters was set up at the Orleans Inn. In a downstairs room, old Cape-Codders, gaunt men with chiseled faces, knotted beards, and weathered hands, mostly fishermen or those who made their livelihood from fishing and had the scars to prove it, gathered. Stanyan stood out like an alien among them, with his hairless chin, polite demeanor, and pinkish complexion—not to mention his wool topcoat, derby, and suit, featuring a vest dangling a gold watch chain from its pocket, as well as a high-collar white shirt and bow-tie. Stanyan was definitely from "off Cape," as the Cape-Codders referred to those not from the region.

The Cape fishermen wore practical uniforms of knee-high rubber boots, ragged knitted turtleneck sweaters, breeches, and heavy dark coats. Their eyes were flinty and their beards ragged and salty. At one time, they could have easily passed as pirates—perhaps, in point of fact, some of their ancestors had been. Their heads were covered with heavy wool caps. Some had long fishing knives stuck in their wide, thick belts or jutting out of the pockets of their pea-coats. Stanyan had a notebook and pencils in his pockets.

They were volunteers come to help the lifesaving crews patrol the beaches searching for bodies. Lifesaving crews had been working day and night recovering bodies. Shortly after his interview with Johnson, fifteen bodies had been recovered.

Now came the gruesome process of identifying them. Cape Cod physician and coroner Dr. Samuel T. Davis, was placed in charge of the operation. Lifesavers reported sighting several bodies in the surf some distance from shore, but they disappeared out of view in the powerful current off shore and could not be recovered. Bodies were taken to funeral homes in Provincetown and Orleans to await identification.

Stanyan and other reporters were recruited to help. He spent the day and night working with a Nauset Beach crew, trudging up and down the bleak coast searching for bodies. By then, wreckage from the steamship also littered the beaches and dunes.

> WRECKAGE 15 MILES SOUTH
>
> Dr. Maurice Richardson of Beacon Street, this city, has been at his Summer home at Wellfleet during the storm, and his story corroborates the early accounts of the loss of the *Portland*, for he saw two of the bodies washed ashore and on them were life preservers marked with the vessel's name. Dr. Richardson was on the first train from Cape Cod which arrived in this city late to-night. To take the train he was obliged to ride fifteen miles. "I saw two of the bodies picked up," said Dr. Richardson. "One was probably that of a deckhand, a man of about twenty. He had on a life preserver marked 'Portland.' The other body was that of a stout woman. She, too, wore a life belt with the steamer's name on it. Wreckage is coming ashore for fifteen miles along the coast." He said that at Orleans the body of a girl of about twenty was found. She had a gold watch and a ring marked "J. G. E." Her watch stopped at 9:17. (*The New York Times*, November 30, 1898)

The frenzy along Nauset Beach was made even worse by beachcombers and scavengers who were hauling away any salvage they could get their hands on, from the ship's battered timbers and railings, to unopened cans of food and other sundries. Stanyan retrieved anything he could find from the surf or sandy beach that he thought might be helpful in identifying the victims of the wreck. Everything was dutifully turned over to Dr. Davis, who slowly pieced together the identity of the victims based on the merest shreds of evidence.

> Large numbers of volunteers, including several persons who had friends or relatives on board the steamer, assisted the surf men in patrolling the beaches. From the tip end of Cape Cod to Monomoy there are ten Government life saving stations, and all the crews have been on duty almost constantly since last Saturday evening,

> when the great blow set in. On Monday and yesterday several bodies were sighted in the surf some distance from shore, but they disappeared from view before they could be secured. (*Sacramento Daily Union*, December 1, 1898)

But even as the names of the victims were beginning to come to light, Stanyan and the other reporters remained stymied–telephone and telegraph wires were still down all over Cape Cod and reporting the names of those whose bodies had so far been identified had to wait until the lines were repaired the next day, if then.

Drained and dispirited by the day's events, Stanyan headed back to his room at the Orleans Inn, but unlike his fellow reporters, who appeared to accept the communications debacle brought about by the storm, Stanyan knew there had to be some way of getting the names back to his editorial offices in Boston. The resourceful solution for him came not as a blinding illumination but, instead, as the tiniest of sounds–a rhythmic click, click, click that he heard coming from a small lighted building down the street from the inn. He had seen the lights on inside the small building and, transfixed, trudged through the snow toward it. As he neared it, he heard the click of a telegraph key.

Stanyan knocked and let himself into the building. Inside was a young man working late sending signals across the ocean to France, using the transatlantic cable. What Stanyan had stumbled on was the Cape Cod office of the French Cable Station.

In 1879, the French company, Compagnie Française du Télégraphe de Paris à New York, built a transatlantic cable from Brest, France, to the island of St. Pierre off the coast of Newfoundland and then on to Cape Cod. It stretched 2,242 nautical miles under the Atlantic to St. Pierre and 827 nautical miles from there to Cape Cod. The transmission station was built in Orleans, near the town's commercial district.

Stanyan convinced the young clerk working in the office that night to send his story and the list of *Portland* victims that had been identified so far to his editorial offices in Boston. Since there was no direct connection from the French Cable Station to Boston, he had the clerk telegraph the first-ever transatlantic news story. The signal went the circuitous route from Orleans to St. Pierre Island, to Brest, France. From there, the signal was sent to London, to Ireland, to Nova Scotia and down the storm-ravaged New England coast to Stanyan's editorial offices at the *Boston Globe*. Fifteen bodies from the steamship *Portland* had been recovered, and some were positively identified. The story ran in the *Boston Globe* on Monday.

STEAMER PORTLAND SURELY LOST

> Evidence of horrifying proportions has washed ashore onto a deserted Truro beach on Cape Cod indicating that the steamship Portland, bound from Boston to Portland, Maine and leaving on November 26 prior to the horrific storm that swept mercilessly across the New England coast was surely lost at sea.
>
> Wellfleet lifesaver Jack Johnson was the first to report the discovery of a life jacket belonging to the Portland along with an unopened can of perishable food stuff that surely came from the missing steamship.
>
> No word of the ship has been heard from anywhere along its regular route.
>
> The managers of the Boston and Portland Steamship Company, stated to-night that there were grave doubts as to the safety of the steamer Portland, which sailed from here Saturday night.
>
> Every harbor between here and Portland on the north shore has been heard from, and one of the south shore, and not in any case has the steamer been seen. The only remaining harbor which she could have reached is Provincetown, on Cape Cod, and news from that port is anxiously awaited. It is still impossible to reach Provincetown by wire.
>
> All the above officers with the exception of the Captain and pilot belong in Portland. The steamer also had two stewardesses, making in all ninety-seven souls on board.
>
> Arrangements have been made with the Government to dispatch the revenue cutters *Dallas*, from this port, and *Woodbury* from Portland to hunt up the steamer.
>
> The *Dallas* will cruise along the south shore as far as Provincetown, while the *Woodbury* will make a long circuit from Portland

> around Cape Ann, and if neither of these steamers succeed in locating the missing vessel there is little hope of her ever being seen again. (Frank Stanyan, *The Boston Globe*, November 29, 1898)

Following the publication of the story by Stanyan, a dark cloud of sorrow and disbelief fell over Boston and Portland, where families had been waiting anxiously for some word of the fate of the *Portland*'s crew and passengers. Despite it being his first front-page story with his byline on it, Frank Stanyan found no exhilaration in its publication. And even though the sinking of the *Portland* and those who had gone down with it remained in the abstract to him—names and numbers—it was as if he, and he alone, had sent back from Cape Cod a lifetime of sorrow and misery in the words he had written and published. It was as if he had, by the telling of the story, somehow made it all come true. He could not shake the feeling that he was somehow responsible for the tragedy.

Other gruesome newspaper reports followed.

9: The Long Trek

There was no easy way for Charlie Ward to get his story back to the *Boston Herald* offices in Boston. Like other Boston reporters, Ward had been sent down to cover the story on Cape Cod. Like his counterpart Frank Stanyan from the *Boston Globe*, Ward had taken the train to Sandwich, then a carriage and sleigh to get to Orleans. But unlike Stanyan, he was not able to have his story transmitted via the trans-Atlantic cable.

Getting another sleigh or carriage was impossible. Even the one he had used to get into Orleans was now put into service helping to clear roads and clean up the damage. The only train service began in Sandwich, and the only coach or carriage service started in East Dennis nearly ten miles away. Ward had only one option. He would have to make the ten-mile trek to East Dennis and take a coach to the train in Sandwich.

> Sandwich, Mass., November 29 (by train)—Communication between Cape Cod and the outer world once more has been established. The first train to run below Buzzards Bay since Saturday left that place at 1:45 o'clock to-day and ran to Sagamore. Here the passengers, mails and baggage were transferred by teams to a point below Sandwich, a distance of three miles. A bad break at Truro prevents the train from leaving Provincetown, but passengers from that place were transferred by teams. The washouts on both sides of Sandwich and that at Truro are being repaired. It is thought that trains will be running after Friday. Sandwich received the first mail last evening and sent one this forenoon. Towns below received the first mail to-day since Saturday. None was sent out from there until this afternoon. (*Brooklyn Daily Eagle*, November 30, 1898)

It was slow going for Ward. The sand dunes were covered in six-foot-high drifts of untouched snow. For as far as he could see along the flat uninterrupted plains, both ahead and behind, there was an endless stretch of white. Scrub pines, ripped out by their roots, were strewn across every bit of roadway. Dirt roads were glazed over with ice. Debris littered every step of the way. It was freezing and there was not a soul in sight. As he trudged on, clapping his hands to keep them warm, breathing out great gusts of cloudy breath, stomping his feet to keep his feet from freezing, he surveyed the bleak totality of the storm's devastation. It was as quiet as death.

Ward trudged in knee-deep snow from Orleans to East Dennis. There he was able to hire a coach to Sandwich to catch the only available rail service back to Boston. It took Ward most of the day to make the long trek on foot. He began at sun up and it was dusk by the time he reached the train station in Sandwich. He had only stopped occasionally to rest and to warm himself with sips of brandy from his pocket flask. The notebook weighed as heavy as a casket in the pocket of his coat, and it might well have.

He gladly boarded the train bound for Boston and arrived late on Tuesday, November 29, 1898. Ward's story ran on the front page the next day.

THE BODIES FOUND

The following constitutes a list of the bodies washed ashore on the outside coast of Cape Cod up to midnight to-night:

At Orleans–E. Dudley Freeman of Yarmouth, Me., a prominent attorney and member of the Governor's Council. (identified by name on inside of watchcase and on paper in pocket;) George W. Delaney, twenty-eight years old of 690 Shawmut Avenue, Boston. (identified by card and documents in pocket.)

At Wellfleet–George Graham, (colored,) porter of the *Portland.*

At Provincetown–William Mosher of Gorham, Me.

On Nauset Beach–Body of a man believed to be John Walton, second engineer of the *Portland.*

At the undertaking rooms of Thomas S. Taylor, Provincetown is the body of a woman about fifty years of age, with large frame and features, iron gray hair, and dark eyes. No clothing was on the body except fragments of underwear. The body was somewhat bruised. This body was picked up near Peaked Hill Bars Life-saving Station.

The body of an unknown colored man lies at the undertaking rooms of Nathaniel Gifford of Provincetown. It is that of a man about thirty years old, probably one of the stewards of the steamer. In a pocket a bunch of stateroom keys was found. This body also was picked up near Peaked Hill Bars.

The body of a woman, 5 feet 9 inches in height with light hair, slightly mixed with gray: blue eyes, weight 160 pounds, lies at J. B. Steele's undertaking establishment, Orleans. The woman was about forty-five or fifty years of age. The body was devoid of clothing when picked up on Nauset Beach by the Nauset Life Saving crew. It is judged that the woman had false teeth, as all the teeth are missing. On a finger was a chased ring, with the words "Forget me not."

Mr. Steele has also the body of a girl, which was found last night off Orleans by John G. Rogers. The girl was not over twenty years of age. She had blue eyes, dark brown hair, light complexion, and a full set of teeth. The body was dressed in underclothing with black corsets and a woolen jacket. On the little finger of the right hand was a ring which had a stone in the center. The stone had been evidently washed away. The side settings of the ring were pearls.

Joseph Mayo undertaker at Orleans has the body of a mulatto girl, twenty years old, weight 115 pounds, height 5 feet 3 inches. It is evidently that of one of the waitresses on the *Portland* and is possibly the daughter of one of the stewardesses. The body was partially dressed and had on a plaid cape. There was an opal ring on the third finger of the left hand. The girl's hair was remarkably long and thick.

Another body at Joseph Mayo's awaiting identification is one found on Monday by W. H. Hopkins. This is the body of a woman about forty-five years of age, 5 feet 3 inches in height, weight about 200 pounds. She had on a black petticoat and a dark dress. A gold watch was found on this body with the monogram "J.G. E." engraved on the case. It is thought the body may be that of Mrs. Jennie Edmunds of 21 Marion Street, East Boston, but the identification is by no means positive. She wore a diamond ring, a plain gold band ring, a diamond horseshoe brooch, and a braided chain with a gold slide.

At Mayo's rooms is also the body of a woman about sixty or sixty five years old. The face was very badly disfigured and the body was almost entirely nude. The only means of identification was a bloodstone ring with the initials "L W G.," followed by the figures 79.

At Orleans a second body of a white is held for identification. It is that of a girl of about twenty. The body was fully dressed.

Also there is the body of a colored waiter, about twenty-five years old. Also at Orleans, a man of twenty, with pompadour-cut hair and dark complexion, and a man of forty-five, light complexion, good clothing. [Description apparently referring to the same man coming from Highland Light says a card marked, "John W., Congress Street, Portland" was found on the clothing.] The descriptions of three bodies at Eastham could not be obtained here tonight. (Charlie Ward, *The Boston Herald*, November 30, 1898)

10: Grief and Joy

Newspaper reports sent from Cape Cod to the Boston, Maine, and New York newspapers brought grief to many and joy to others. For some there was only grief:

> The bodies of George W. Delaney of Boston and Mrs. C.M. Mitchell of North Easton, Mass., have been positively identified among those that have been washed ashore.

> Orleans, Mass., December 5–The body of James W. Flower of Lewiston, Me., one of the victims of the lost steamship *Portland* is at an undertaker's here. The body has been positively identified as that of Mr. Flower.

> Brockton, Mass., December 5–At least two of the victims of the ill-fated *Portland* were residents of this city. C. F. Wilson was up to Saturday a clerk at Hotel Metropolitan in this city. He was about 25 years of age. He left this city Saturday to go to Maine. He had a brother who was a clerk in Crawford House, Boston. H. Carter, a fireman, was for some time employed in this city and left here to take his position on the *Portland*. Louis Metcalf of Campbello, the south end of this city, also is supposed to be among the victims. His wife is sick at Auburn, Me., and he left Saturday to join her. He has not been heard from since.

The dead continued to wash ashore. The body of a young man washed up on the Wellfleet Beach and was taken to Provincetown for identification. He was later identified as George Kenniston Jr. from Boothbay, Maine, a student at Bowdoin College. He had in his possession a leather-bound journal. The waters had washed away everything written inside it.

The body of a middle aged African American man washed ashore along the Truro coast. There was no identification on him except for a gold cross around his neck. It would take the coroner's office two weeks to identify him as Eben Heuston, chief steward of the steamship *Portland*.

The body of a girl was found off the Orleans coast. The girl was not over twenty years of age. She had blue eyes, dark brown hair, and a light complexion. There was a length of heavy rope tied around her waist. Emily Cobb's body was taken by train back to her home in Portland where her mother had her buried beside her father. The pastor at the First Parish Church in Portland eulogized: "Her gift of song became the object of her ambition. There was a promise of great joy and worth in such a nature and with such a gift."

The bodies of Captain Hollis Blanchard and Harry Rollinson were never recovered.

> Former Maine Sen. E. Dudley Freeman left a grieving wife and young children. Charles H. Thompson, who ran the Thompson grocery store in the Woodfords section of Portland, was on board with his wife Susan and 3-year-old daughter Gladys. Oren Hooper, who had a furniture store in downtown Portland, perished with his 13-year-old son. There were several local school teachers, including 33-year-old Sophie B. Holmes. George Kenniston was a 20-year-old student at Bowdoin and the youngest member of a prominent Boothbay family. (John Richardson, "Ship of Doom," Blethen Maine Newspapers, 1998)

For others, news brought cries of joy and relief.

> Salem, Mass., December 2–Word was received here from Ipswich this morning that Mr. and Mrs. Julian A. Fogg who were reported on the steamship *Portland*, did not take the steamer on Saturday, being afraid of the storm when they arrived in Boston and deciding to postpone their trip.

> Portland, Me., December 3–It was learned this morning that L.F. Strout, first pilot, and E.B. Deering, mate, previously reported aboard the steamer *Portland*, were not with her when she was wrecked off Cape Cod.

For Captain Bill Thomas, skipper of the fishing boat, *Maud S.*, who had last seen the *Portland* steaming off the coast of Thacher's Island on the night of the storm there came only sweet relief.

> Portland, Maine, December 6–It was learned here today that Mrs. Hope Thomas, beloved wife of Captain William Thomas who had been scheduled to return home to Portland on the doomed *Portland* was not onboard, having decided to wait out the storm with relatives in Boston.

For Margaret Ann Heuston, the wife of *Portland* chief steward Eben Heuston, the news of her husband's death was devastating. Judith Cobb, the widowed mother of Emily Cobb, was overwhelmed with grief. George Kenniston Sr. began making funeral arrangements for his son as did the family of former senator E. Dudley Freeman. The body of Captain Hollis Blanchard was not found, but his wife, Emma, steeled herself in the heartbreaking realization that he had gone down with his ship. All of them, lost at sea.

> Loss of the Steamer *Portland*
> On the twenty-seventh of November
> In the year of ninety-eight,
> A northeast blizzard swept the sea,
> Death following in its wake
> And many good ships floundered,
> Or were stranded on the coast;
> And naught but broken timbers,
> To show how they were lost.
> The clouds were dark and threatening,
> The "glass" was falling low;
> The weather bureau signals
> Foretold a stronger blow,
> When the steamer *Portland* left her dock,
> Proceeded down the bay,
> With over one hundred souls on board.
> O where! O where are they?
>
> (Frederic R. Eldridge, Chatham, Massachusetts, 1899)

SINKING OF THE STEAMER PORTLAND.

Every Soul on Board Has Undoubtedly Been Lost. The Exact Number Not Known, but is Estimated at Over One Hundred. Not a Fragment of a Lifeboat or Life craft Among the Debris From the Ill-Fated Vessel Which Has Been Washed Ashore

PROVINCETOWN (Mass.), Nov. 30.–The steamship *Portland,* plying between Boston and Portland, was swallowed by the sea in last Sunday's storm. . . . Of all the ship's company of over 100 souls, not one survived, and the story of the disaster will never be told. . . . Nearly every sea captain and mariner on the Cape, on being asked for an opinion, states that everyone on board undoubtedly was lost. (*Sacramento Daily Union*, December 1, 1898)

11: Conflicting Reports

Because the manifest listing all passengers on board the *Portland* had gone down with the ship, there were conflicting reports on the exact number of passengers that were lost. The *Boston Herald* reported more than 200 lives had been lost, while a story in the *New York Times*, based on reports from the Portland Steamship Company disputed that number.

150 LIVES LOST
WITH THE PORTLAND
Wreckage With Many Bodies is Drifting South
Number of Bodies Recovered
Total Number of Lives Lost Off New England Will Reach Over 200

Orleans, Mass., December–The following dispatch was received from Orleans, Mass., this forenoon:

Five bodies have been recovered here from the *Portland.* The vessel was wrecked just north of Cape Cod. The wreckage with a great many bodies is drifting south, possibly as far as Nantucket.

It is thought that the northeast gale to-day will bring the bodies in. The entire Cape is patrolled. The body of Dudley Freeman is the only one identified as yet. This body will be shipped to Portland tomorrow.

Dead bodies from the wreck came ashore at Nauset all day yesterday. Life preservers marked Portland were on them. Other bodies have also been picked up at Wellfleet, Truro, Provincetown and at the Orleans lifesaving station. Up to last night thirty-six bodies, both men and women, have been reported found along the shore in this vicinity. The life savers and coast patrol have been on duty constantly and hundreds are assisting in the work of rescuing the bodies from the terrific surf. The sea is still running so high that it is only with great difficulty that bodies can be landed. Many of them are frozen stiff. And in such condition that identification is very difficult. Wreckage of all kinds is drifting in. Fragments of boats marked Portland, portions of the cargo and splinters from the steamer are strewn all along the shore for fifteen miles. (*The Boston Herald*, November 30, 1898)

PROVINCETOWN, Mass., Nov. 30.–The exact number of persons who were carried away from Boston by the Portland will probably never be known, as no list of passengers was retained on shore when the vessel left last Saturday. Many estimates of the number onboard have been made, but the estimates seldom agree.

C. F. Williams, Boston agent of the Portland Steamship Company, who arrived here on the tug William H. Smith last night, places the total number of persons on the steamer at 100 or possibly 105. This estimate, however, is generally regarded as rather small. It has been stated that the number was as high as 155, but Mr. Williams denies that so many sailed on the Portland. It is probable that 120, including passengers and crew, is near the correct number. (*New York Times*, December 1, 1898)

And there were other newspaper reports:

> MESSAGE FROM CAPT. BLANCHARD?
>
> Ostensible Appeal for Help Found at Nantasket Beach.
>
> NANTASKET BEACH, Mass., Dec. 6.–A flask containing a message purporting to have come from Capt. Blanchard of the steamer *Portland*, was picked up on the beach nearly opposite Whitehead this afternoon. The message reads: Help! On board the Portland. We are sinking. Upper works gone. Two miles off Highland Light. Time, 7:30 Sunday. (*The Boston Globe*, December 7, 1898)

There was no credence as to the authenticity of the reported message from Blanchard.

12: Vigil

Many of the bodies recovered from the *Portland* wore wristwatches that had stopped at 9:15. According to Andrew Toppan, author of *Haze Gray & Underway*,

> It is unclear, however, if this indicates the ship was lost at 9:15 AM, or at 9:15 PM. Although there are several reports of the ship being sighted, afloat, between 9 AM and 10:30 AM that day, the exact times of those sightings are not known. If any of those sightings took place after 9:15 AM, then the ship must have survived until 9:15 PM that day. . . . However, *Portland* would not have carried enough fuel to remain at sea, in storm conditions, for over 24 hours. She could have burned furnishings, interior bulkheads, and other wooden materials to keep the boilers running, but the quantity of this material washed ashore tends to indicate this action was not taken.
>
> Also, it is highly questionable whether she could have held together for 24 hours, given the terrible sea conditions. Still, the fact that major debris did not begin to wash ashore until 9:30 PM

> suggests that *Portland* had survived into the night–surely, if she had been wrecked at 9:15 AM, debris would have been washed ashore in the morning. Because the exact time of the final sightings cannot be firmly established, it is impossible to conclusively determine the exact time of *Portland*'s loss–either 9:15 AM, or 9:15 PM, on Sunday, November 27, 1898.

For weeks after the sinking of the *Portland*, a vigil was kept along the Cape Cod beaches by lifesavers, volunteers, family, and friends, but no further victims were found. Thirty-six bodies were ultimately recovered from the sea. The last body from the steamer washed ashore on December 6, 1898.

For months, wreckage from the sunken steamer washed up on the Cape beaches. Slowly, one by one, sorrowful and grieving family and friends left the Cape. In the midst of their sorrow they all left with one question on their lips: Why? Why did Captain Blanchard leave port that night in the face of this horrific storm? There were so many unanswered questions. There would be for many years to come.

The Portland Steamship Company officials claimed that Blanchard had been contacted by telephone and explicitly told to remain docked at Indian Wharf in Boston to wait out the storm. Others contended that Blanchard was never contacted by the steamship company and if he had, he was ordered to set for home that evening. No one will ever know the answer and all that remains is speculation.

Following the wreck of the *Portland* a sensational full marine court investigated the circumstances leading up to the loss at sea to determine whether Blanchard was at fault or not. There was testimony given both damning Blanchard as well as absolving him.

13: Liscomb's Statement

The Portland Steamship Company general manager John Liscomb placed responsibility for the disaster on Captain Blanchard. He told the press, "It was clearly a case of bad judgment and disobedience of orders on the part of Capt. Blanchard. . . . Our motto has always been to err on the safe side and, in fact, I have heard the people have sometimes referred to us as the 'Old Granny' line, because we would not sail in threatening weather. We never allow a boat to go out in a gale or a snow storm and if they get caught after they are getting out to sea they are instructed to return. . . . He [Captain Blanchard] was evidently convinced that the approaching storm was not going to be very severe, perhaps founding that belief on the predication which I believe was in Saturday's Government report that the wind on Sunday would be northwest."

Despite the loss of life, the steamship company sought to have its insurance liabilities associated with the wreck of the *Portland* limited.

WRECK OF THE PORTLAND

The Vessel's Owners Ask a Limitation
Of Their Liability

> Portland, Me., Dec. 26–Attorneys for the Portland Steamship Company have filed in the United States District Court a petition for the limitations of liabilities in the loss of the steamer *Portland*. . . . The petitioners also state that the loss of life and property occurred without the knowledge of the petitioners, and occurred by the act of God and by perils of the sea and without fault on their part. (*New York Times*, December 27, 1898)

14: Williams's Letter

A letter written on December 3, 1898, by Charles Williams, the Boston-based agent for the Portland Steamship Company, was not entered into testimony during the hearing. The letter was written to

John Liscomb, the steamship company's general manager in Portland, Maine, and confirmed the conversation he had with Captain Blanchard.

According to the contents of the letter:

> I told George Barton, the watchman, to watch the ring of telephone sure about nine o'clock for you had sent word for Capt. Blanchard to wait till 9 for weather report, but Capt. Blanchard would not for he was bound to go on time and that you would be wild to hear he had not waited. George Barton will swear I told him that and it only goes to show that I said to Capt. Blanchard all that a man could say to follow out your request for him to wait.

The letter belonged to seventy-six-year-old Arthur Liscomb of Arundel, Maine, the great-grandson of John Liscomb. It was not brought to light until more than a hundred years after the tragic *Portland* disaster. The letter had been tucked away in a scrapbook that belonged to John Liscomb that had been kept in the family for two generations.

According to Arthur Liscomb, the scrapbook was kept on a book shelf in his mother's home for years, until she died in 1994. He discovered the scrapbook and the letter while cleaning out her house. Liscomb subsequently brought the letter to the attention of Portland Harbor Museum officials in Spring Point, Maine. The museum's curator, Ned Allen, called the discovery of the Williams letter "a significant piece of evidence on this issue."

Allen went on to say, "Since it was written by Mr. Williams, it's still telling his side. But it certainly doesn't look like the sort of thing one would write after the fact, kind of covering up. It's so emotional. These people obviously just went through a very rough time in their life. This horrible disaster happened on their watch."

"It was always told to me that [Blanchard] was more or less to blame for the sinking of it," the elderly Arthur Liscomb said.

15: Rousmaniere's Book

In his book *After the Storm: True Stories of Disaster and Recovery at Sea* (2002), John Rousmaniere writes, "Hollis Blanchard was one of the most notorious figures in New England history–a Yankee ogre up there with Lizzie Borden and the witch hangers of Salem. He had no admirers after November 1898, once it was believed that he had willfully sailed out into a double storm that had been perfectly predicted by the U.S. Weather bureau."

According to Rousmaniere, the U.S. Weather Service and the Steamship company were more to blame for the disaster than Blanchard. The Weather Service, because it did not accurately inform Blanchard of the severity of the storm bearing down on New England, was also culpable.

Rousmaniere also claimed

> When J.W. Smith, head of the Weather Bureau's Boston office wrote in the bureau's magazine, the *Monthly Weather Review*, that the *Portland* had been "fully warned by the Weather bureau," he did not specify what "fully" meant. There is good reason to believe that it did not include the small low racing up the Atlantic coast or the double storm. . . . Very likely, until the violent upheavals of Saturday night, neither the bureau or Blanchard believed that the problem was anything other than a single gale coming in from the west.

In regard to the Steamship Company's complicity in the disaster, according once again to Rousmaniere: "The evidence at hand indicates not only that the Portland Steamship Company was in a period of transition on late November, but that on November 26 it was in organizational chaos."

John Liscomb, the general manager, had only been on the job a month before the tragedy and although at first he said he had ordered Blanchard to stay in port that night, in later legal proceedings, he testified that he had not ordered Blanchard to do anything but merely "informed" him of the potential dangers. "Thus the line declined to exercise its authority to keep the Portland tied up at India Wharf."

Based on his research, Rousmaniere concluded, "In short, one of the worst disasters in New England's history occurred for banal reasons having to do with bad luck, bad timing and misunderstandings." He cited the National Weather Bureau forecast from Washington, D.C., issued on November 26: "For Massachusetts, Rhode Island and Connecticut heavy snow was predicted with clearing on Sunday followed by much colder weather."

16: Damning Accounts

Of captain brave, he was the best,
To my aye storm a-long!
But now he's gone and is at rest;
Aye, aye, aye, Mister Storm a-long.

(Sea chanty, 1840)

According to damning newspaper accounts following the disaster, Blanchard had taken chances that, "No man in his position had a right to take." An editorial published in the *Washington Evening Star* stated, "The *Portland* captain ignored the official warnings . . . and direct orders of his superior to keep in port. . . . He carried with him to death over 100 people who had no knowledge, presumably, of the desperate chances which he was taking."

In a front-page news story that appeared in the *Portland Evening Express*, published shortly after the disaster, Captain C. H. Leighton told reporters that he had directly questioned Blanchard about his decision to leave Boston that evening in the face of the ominous weather forecast and Blanchard responded cavalierly, "We may have a good chance."

SAILED CONTRARY TO ORDERS

> Portland, Me., Nov. 29–Ever since the *Portland* has been missing there have been inquires as to why the steamer should leave. . . . It is probable that Capt. Blanchard would have been severely reprimanded . . . even if he had been able to come through all right, as he had left in the face of explicit directions to the contrary. It is probable that he thought the storm was not likely to strike as it did by several hours. (*New York Times*, November 30, 1898)

Another article, published in *The Boston Evening Transcript*, suggested that it might have been Captain Dennison's telephone call that sealed the fate of the ship. Disappointed that the younger and less experienced Dennison had been given command of the newer *Bay State* steamship over him, Blanchard may have been trying to prove his worth and skill to Dennison and the company.

17: Defense of Captain Blanchard

Other sea captains came to the defense of Blanchard at the hearing, claiming that that the waters were calm when Blanchard left Boston Harbor and that he had every right to make the decision to leave. According to many of the captains who did come to his defense, the decision to leave port under any condition was far too complex to be ascertained by "self-appointed experts ashore." They felt that Captain Blanchard was being unjustly blamed for the sinking of his ship and was unable to defend himself.

At least twelve of Blanchard's contemporary sea captains testified at the trial that they also would have set sail, given the scant information available at the time. However, some who knew Hollis Blanchard also stated that he was a forceful captain who was bound to keep to schedules. In an interview in Portland's *Daily Eastern Argus*, Captain John W. Craig, a side-wheeler captain who had frequently sailed with Blanchard, said, "A cooler man I never knew and

I doubt if there was his superior on any boat that plies the Maine coast. He was honest, faithful, and fearless. I have never seen in all my seafaring experience a man who could run a boat better in a storm or calm. But when he was pilot with me on the *Tremont* he never wanted to admit that the weather was bad. He couldn't seem to see bad weather and didn't like to talk about it. Sometimes I would ask him if he did not think that it was blowing up a storm, or if the prospect was not rather dubious, but he would seldom admit it."

Blanchard had once confided in Boston weather forecaster E. B. Rideout that he had been disciplined by his superiors for being too cautious in light of the increasing competition from the railroads that also transported passengers and freight from Boston to Portland.

The wreck of the *Portland* weighed heavily on Hollis Blanchard's son, Fred, and he refused to discuss it with anyone. His wife, Lucy, was very outspoken in defense of her father-in-law, whom she felt was unjustly portrayed as being reckless and negligent. She claimed that years after the disaster an employee of the Portland Steamship Company had told her that her father-in-law was pressured by company officials into taking the *Portland* out to sea against his better judgment. According to Lucy Blanchard, the employee said Captain Blanchard was told, "There are a lot of important people who expect to be in Portland tomorrow morning and we don't want to disappoint them."

Grace Blanchard, Captain Blanchard's granddaughter, said that her grandfather had been ordered by his superiors at the Portland Steamship Company to sail on the evening of November 26, 1898. This story was told to her by her father, Fred, who lived in Boston at the time of the *Portland* catastrophe and who spoke with the captain that afternoon. Miss Blanchard stated that when her father asked if it was necessary to sail that evening, Hollis Blanchard replied, "I have my orders to sail, and I am going!"

In May 1899, the United States District Court, District of Maine in Admiralty ruled that the sinking of the *Portland* was "an act of God," and not the fault of Captain Blanchard's poor decision making. The *Portland* was valued at $250,000 at the time of the disaster but was only insured for $60,000. The company was able to recoup a portion of its overall investment; however, the families of the ship's passengers had no recourse in the courts.

> Under an act of June 7, 1871, Congress authorized the office of "shipping commissioner" in federal circuit courts whose jurisdictions included seaports or customs ports of entry. Among their duties, commissioners collected the wages due to deceased seamen from the vessel's master or owners and turned the money over to the court for delivery to the legal heirs. Unclaimed monies would eventually be turned in to the U.S. Treasury. Claimed funds would be paid only after the court was satisfied that the proper person was getting the money. To that end, these case files provide extensive family information through preprinted forms or affidavits or both. (Walter V. Hickey, "The Final Voyage of the *Portland*," *Prologue Magazine*, 2006)

When the *Portland* went down on November 27, 1898, the records on board went down with her. Some of the crew on board the *Portland* were identified from the various wage papers filed with the shipping commissioner and the court. Among the families of crew members who filed documentation was John C. Whitten, a watchman on the ship who was survived by a wife and four children. He was owed $35 for one month's wages. According to the paperwork, his widow filed the claim "on behalf of herself and her children, ages 6 to 12."

Ellen Johnston of Portland, Maine, lost both her husband, Arthur, and her sister, Carrie Harris. Johnston filed a claim for her husband's wages on December 19, 1898. According to the documentation, "She stated that her husband had been born at Moose River, Annapolis County, Nova Scotia, on April 2, 1848. His parents were both deceased, and his survivors included three children, Archibald, Harriet, and Josephine. The 1900 census of Portland, Maine, showed Helen [*sic*] with her children and revealed that they were black, all born in Nova Scotia."

Johnston also filed a claim for the wages of her sister. Carrie Harris was a widow and had no children. "Her siblings included one brother, John Sibley, and two sisters, Cellia Sibley, and Ellen Johnson [*sic*]. In January of 1899, both John and Cellia Sibley sent affidavits from Nova Scotia to the court in Portland, asking that their share of Carrie's wages be paid to their sister, Ellen Johnston."

The sinking of the steamship led to significant changes within the maritime industry. Most significantly, it hastened the demise of wooden, side-paddle steamships. Within the decade, steel-hulled, propeller-driven steamers were introduced as the main source of transportation along the treacherous New England coast. Following the *Portland* disaster, all passenger vessels were equipped with wireless transmitters and copies of passenger ship manifests were kept not only on board but in safe-keeping ashore. All these safety precautions and new regulations came too late for the ill-fated *Portland*.

The steamship company reorganized in order to avoid liabilities associated with the death of all passengers and crew.

> NEW STEAMSHIP COMPANY
>
> Said to be Formed to Protect the Portland Steamship Company Against Suits for Damages.
>
> PORTLAND, Me., March 11–There were filed in the office of the Register of Deeds this forenoon papers pertaining to the organization of a new steamship corporation, to be known as the Portland Consolidated Steamship Company. Its capital stock is $500,000, and the Directors are the present officers of the Portland Steamship Company. (*New York Times*, March 12, 1899)

> The disaster also marked a turning point in the way steamship companies did business. Following the *Portland*'s loss, companies stopped using shallow-hulled side-wheelers, switching instead to deeper-draft, propeller-driven ships that were more stable in rough seas. Companies also made sure they kept copies of passenger lists on shore. The *Portland* carried the only copy of its final passenger list to the bottom, making it difficult to identify–or even tally–those lost. (Peter N. Spotts, "Scientists Find the '*Titanic* of New England,'" *Christian Science Monitor*, August 30, 2002)

Despite the reorganization, the steamship company was not able to stay afloat after the *Portland* disaster and in 1901 it was bought out by Eastern Steamship Company, ending its fifty-seven-year service between Boston and Portland.

18: New England Shipwrecks

> They that go down to the sea in ships, that do business in great waters . . .
>
> –Psalms, 107:23–30

Although the sinking of the *Portland* was the most tragic and sensational maritime disaster along the New England coast, it was not, of course, the only one. The New England coast had long been the site of some of the country's worst wrecks. According to an article by Amy Woods, "Shipwrecks along the New England Coasts," appearing in volume 31 of *New England Magazine* in 1904, "In the ten years commencing June 1st 1879 and ending June 1st 1889 the record shows a yearly average of 1,919 wrecks and 3,535 lives lost on or near the coasts and on the rivers of the United States."

From Maine to Rhode Island, the sandy bottom is littered with the horrifying wrecks of merchant and passenger ships, their crews and passengers all lost. It would be impossible to write about all the many ships that wrecked along the New England coast in its tumultuous waters, but there are three that exemplify tragedies at sea. They include the wreck of the *Asia* in 1898, whose first mate made a solemn vow to the captain of the ship about saving the captain's young daughter; the sinking of the *Royal Tar* in 1836, because of its unique cargo; and the sinking of a Cuttyhunk life-saving rescue boat. Although these events were not as widely publicized as the sinking of the *Portland*, they are nonetheless worth remembering because of the tragic circumstances surrounding them and the crew and passengers who all went "down to the sea in ships."

Although the number of victims on board the *Asia* in February 1898 could not compare to the lives lost on board the *Portland*, the circumstances surrounding the sinking of the *Asia* were no less heartbreaking. On board the ship, besides a crew of six men, was Captain Daniel Dakin, his wife Eleanor, and his eleven-year-old daughter, Lena. The voyage was intended to be Captain Dakin's last. He had enjoyed a prosperous career and was returning to Boston to retire. As the *Asia* neared the shoals off Nantucket Island it was struck by a raging winter storm and it slammed bow first into the sandy bottom of one of hidden shoals. The relentless New England storm pounded the stranded vessel until it broke apart. The ship's first mate vowed to the captain that he wouldn't let Lena drown. He tied a rope around her waist and tied the other end of it around his. The waves washed everyone into the frigid sea. Only three crew members managed to survive. It was not until weeks later that the bodies of first mate and young Lena Dakin were found, frozen to death, in the icy waters off the coast of Nantucket Island, still tied together. The first mate had kept his promise, the young girl had not drowned in the icy sea but had frozen to death tied to the young man who had promised to save her.

In one of the most bizarre shipwrecks in New England lore, the sinking of the 164-foot steamer, *Royal Tar*, in October 1836 the ferocious sea took the lives of a menagerie of circus animals, including a giant elephant named Mogul. The *Royal Tar* sailed out of Saint John Harbor in New Brunswick on Friday, October 21, 1836, after a successful touring season in the northern provinces, with an entire circus aboard, including performers, animals, and a brass band, all bound for Maine. On board were show horses, camels, two lions, a Royal Bengal tiger, several pelicans, and the elephant, Mogul.

Along its route the ship's boiler exploded, setting fire to the main beam next to the elephant stall where Mogul was kept. The flames moved rapidly through the ship, engulfing everything in the lower decks. Pandemonium broke out aboard the *Royal Tar*. The winds blew hard across the flaming decks. Circus performers, crew, and animals surged across the deck of the burning vessel, billows of smoke engulfing them. The *Royal Tar* was burning out of control. Mogul, the elephant, terrified by the raging flames, rose on his hind quarters and then leaped off the side of the ship into the water. The body of Mogul was later found washed ashore several miles away. Although all the passengers and crew were saved, all the circus animals drowned.

In the tragic tales of New England shipwrecks, it is sometimes not the captain, crew, and passengers who fall victim to the fury of the sea. The story of the *Aquatic*, wrecked off the coast of Cuttyhunk Island on February 24, 1893, is such a story.

The ship ran aground along the island's rocky and storm-swept coast. Flares on board the ship were fired to alert rescuers. Rescuers at the Cuttyhunk Lighthouse first spotted the grounded ship. At Boathouse No. 43, there were six Humane Society rescuers. It was nine o'clock at night and the winter gale was still pounding the coast. The rescuers were advised to wait until morning before trying to make a rescue attempt.

Regardless of the warnings, the crew of Boathouse No. 43 launched their lifeboat into the pounding, freezing surf. From shore it looked for a time that the crew just might make the rescue, as they rowed their boat over and through the cresting waves. Soon they were within fifty yards of the wreck, with the cold, powerful current sweeping them even closer to their destination. A mere twenty yards from the wreck, the crew decided to anchor and run a line to the *Aquatic*. A huge wave crashed into the lifeboat and capsized it. The boat went over and the crew with it. Two brothers on board the rescue boat, Fred and Tim Akin, were able to cling to the capsized boat, but the current was too strong and the water too cold. Both men drowned, as the horrified crew of the *Aquatic* watched. Although the crew of the Aquatic were all ultimately saved by another rescue boat, the crew of Boathouse No. 43 were lost.

19: Yearly Commemorative

Beginning in 1908, a handful of surviving relatives of those lost on board the *Portland* gathered yearly on Boston's India Wharf to commemorate the anniversary of the sinking of the steamship.

George Kenniston Sr. and his wife attended the anniversary ceremonies every year in honor of their youngest son, George Kenniston Jr. Kenniston started a scholarship at Bowdoin College in memory of his son and continued to attend the yearly ceremonies in Boston commemorating the sinking of the *Portland* until his death in 1929.

Margaret Ann Heuston religiously attended the Boston observance. Every year she would toss a wreath of flowers into the harbor in memory of her husband, Eben. There were sixty-five members of the African American community from Munjoy Hill who died on board the *Portland.* The Abyssinian Congregational Church in Portland lost nineteen of its congregation. Every year the church held a service in memory of those on board the *Portland.* Less than a decade after the tragedy the church had a mere dozen members. The church ultimately closed. Margaret never remarried. She continued to live on Munjoy Hill where she raised her children by herself and worked as a school teacher. She died in 1938. She was eighty-two years old.

> In November 1898, the 281-foot SS *Portland* sank off the coast of Cape Cod in what has become known as the "Great Portland Gale." The loss of nearly twenty Abyssinian supporters and members weighed heavily on the church and the community. The Abyssinian went into decline by the late 1890s and in 1917 it closed. (Maureen Elgersman Lee, *Black Bangor: African Americans in a Maine Community*, 2005)

According to Portland writer and historian Herbert Adams, Portland's tight-knit African American community on Munjoy Hill was hit especially hard by the sinking of the *Portland.* Many of the ship's sixty-five crew members were African American men from Portland. Many had families there and were leaders in the Abyssinian Church. The church held services and paid tribute to the members who had been lost including two church trustees and seventeen other members.

A newspaper account reported that the words of Rev. Theobold A. Smith, who spoke at the service, "brought out those nearer to them into tears and but few at times could be seen with dry eyes."

Adams, maintained that the Abyssinian Church was also a victim of the sinking of the *Portland.* "The blow was so heavy that it quite literally wiped out the active supporters . . . and within 12 years the congregation was down to about seven active members," he said.

> The tragic sinking of the SS *Portland*, a steamship that became a watery grave off Cape Cod for nearly 200 passengers and crew members 119 years ago, was commemorated Friday evening at the Maine Historical Society. Members of the society and the Abyssinian Meeting House Restoration Project held a ceremony in the Brown Research Library, where they read the names of known passengers and crew members, and rang a maritime bell after each name, said Kate McBrien, the society's chief curator. . . . The loss was so devastating to the congregation, its membership declined thereafter and the church closed in 1917." (Kelley Bouchard, "Nearly 200 Victims of the Sinking of SS *Portland* Are Remembered," *Portland Press Herald*, 2017).

Judith Cobb attended the ceremony at India Wharf in Boston for several years in memory of her only daughter. Because of her failing health, she was unable to attend every year. Associate pastor John Carroll Perkins held a memorial service for Emily Cobb at the First Parish Church in Portland every year. Judith Cobb passed away in 1918 during the influenza epidemic that swept across the country.

A memorial marker was erected and dedicated to the loss of the *Portland* on November 27, 1948, by Edward Rowe Snow and the surviving members of the Portland Associates. The marker is located in front of the building adjacent to the Cape Cod Highland Lighthouse in Truro, Massachusetts. The inscription on the stone marker reads: "On November 27, 1898 the steamer *Portland* with 176 persons aboard sank with no survivors about seven miles out to sea from this station at Cape Cod Light."

In 1948, on the fiftieth anniversary of the tragic event, the group disbanded. The elderly among them were in failing health and the young had no memories of the event. According to New England historian and author Edward Rowe Snow, who spoke at the last gathering, "the 50th anniversary seemed to be a good time to stop."

Massachusetts: Last Voyage

On days when the fog lies still and heavy over the harbors, when the damp beads the dock lines and the only sound is the creak of fenders against pilings, New England's fishermen can still strike up an argument over the loss of the steamer *Portland*. Her sinking, with the loss of all hands, is New England's most famous shipwreck, and the 1898 gale in which she went down is still known, from Nantucket to Bangor, as "the Portland gale. . . ."

Last week a handful of surviving relatives gathered, as they have regularly since 1908, to commemorate the anniversary. Sitting on upended fish boxes in the chill, barn like steamer shed on Boston's India Wharf, they listened as Historian Edward Rowe Snow recounted the oft-told tale of the Portland's sinking.

In 1945, the hull was found. It was lying, sanded in, among huge boulders some four miles off shore and 145 feet down. Her superstructure swept away, she had gone down like a sounding lead in deep water.

Last week, on the chill wharf, the surviving relatives heard the roll call of the Portland's dead for the last time. As each name was called, survivors threw flowers on the ebbing tide. A woman played Rock of Ages on a zither. It was the last meeting. The old were ailing, the young had no memories. Said Historian Snow: "After all, you've got to stop some time, and the 50th anniversary seemed to be a good time to stop." (*Time*, December 6, 1948)

20: Haunting Loss

The sea does not reward those who are too anxious, too greedy, or too impatient. One should lie empty, open, choiceless as a beach–waiting for a gift from the sea.

–Anne Morrow Lindbergh

The events that led up to the *Portland*'s demise and the actual location where it descended to the seafloor have haunted the citizens of

Portland and Boston for over ninety years. This sense of mystery led explorers to search for the vessel throughout the twentieth century. In 1899, the *Boston Globe* funded an undertaking to find the wreck on Peaked Hill Bar. Searchers dragged a cable between two boats along the sea floor looking for wreckage, but found nothing.

For decades, controversy reigned regarding the location of the doomed ship. In 1945, Edward Rowe Snow commissioned a deep-sea expedition headed by diver Al George from Malden, Massachusetts, to explore what Snow believed were the remains of the *Portland*, located some seven and a half miles from Provincetown, Cape Cod's Race Point Coast Guard Station.

George reportedly discovered the wreck in 144 feet of water. He described the hull of the steamship as buried almost entirely in the sand lying among huge boulders some four miles off shore. Her superstructure was swept away. George was able to bring up a stateroom key from the vessel that was clearly marked *Portland.* Not far from it were the remains of the schooner *Addie E. Snow.* Although Snow supported this claim for many years in his various publications, other maritime historians doubted its accuracy.

Snow recorded the affidavit of George in his book *Strange Tales from Nova Scotia to Cape Hatteras.* According to the affidavit, George found the site by traveling to a location discovered by Captain Charles G. Carver of Rockland, Maine. In 1924, Carver reportedly brought up several items thought to be from the *Portland* in his fishing nets at this particular site. The site was identified as, "Highland Light bears 175 degrees true at a distance of 4.5 miles; the Pilgrim Monument, 6.25 miles away has a bearing of 210 degrees; Race Point Coast Guard Station, bearing 255 degrees, is seven miles distant."

According to George, recovering any artifacts of the sunken *Portland* would be costly and nearly impossible, given the condition of the wreck: It was severely embedded in sand and its remains were extensively distributed along the ocean floor, virtually spread out in bits and pieces.

21: Haunting Loss Found

John Fish and his associates began to search for the *Portland* in 1981. Fish specialized in underwater search and recovery operations and was the vice president of the Cape Cod–based company American Underwater Search and Survey (AUSS). He was also a noted maritime history author and sonar expert.

By 1989, Fish and his colleagues had spent thousands of hours at sea combing Massachusetts Bay looking for the wreck. Working with oceanographers from the Woods Hole Oceanographic Institution, Fish recorded what washed up on the Cape Cod shores and according to reports, "analyzed the data with computer software that factored in possible wind and tidal drift. They plotted hourly positions for the wreckage back to a possible sinking site."

The 1989 discovery of the *Portland* was the culmination of nearly a decade of work by Fish and others.

> During the search for the *Portland*, Fish and his colleagues at HMGNE spent thousands of hours at sea combing Massachusetts Bay looking for the vessel. In consultation with a physical oceanographer at the Woods Hole Oceanographic Institution, they recorded what washed up on the Cape Cod shores and analyzed the data with computer software that factored in possible wind and tidal drift. They plotted hourly positions for the wreckage back to a possible sinking site (based on the evidence of stopped watches found on the bodies of recovered *Portland* victims). Their 1989 discovery of the *Portland* was the culmination of nearly a decade of effort. (NOAA, National Oceanic and Atmosphere Administration, U.S. Department of Commerce, 2010)

According to an April 1989 article in the *Orlando Sentinel*, Fish claimed, "There's no doubt; we've got it." The wreck was found in 300 feet of water off the coast of Cape Ann. "The find is particularly significant, not only because the *Portland* was one of the worst maritime disasters in the Northeast, but because it remained shrouded in mystery," Paul Johnston of the Smithsonian Institute said.

22: Ship Positively Identified

In the summer of 2002 the Stellwagen Bank National Marine Sanctuary, an agency of the United States Department of Commerce, and the National Undersea Research Center at the University of Connecticut brought back images that positively identified the ship sitting at the bottom of the ocean in approximately 400 feet of water within the 800-plus-mile Stellwagen Marine Sanctuary.

> It wasn't until 2002 that the submerged wreck of the *Portland* was located. Using data generated by the American Underwater Search and Survey a few years earlier, a survey vessel located the soggy remains of the Portland. High resolution images of the wreck were taken to confirm its identity. Debris fields, akin to the *Titanic*'s, were found. . . . Located seven miles off Massachusetts, the site of the sunken *Portland* is a lonely, cold grave for what little remains of the ship and its 192 passengers and crew. Other wrecks, victims of earlier New England gales, rest nearby. Located well within U.S. waters, the *Portland* wreck was added to the National Register of Historic Places in 2005. (Louis Varricchio, "The Wreck of the S. S. *Portland*: The *Titanic* of New England," *The New York Sun*, 2012)

The Stellwagen Marine Sanctuary is one of thirteen national marine sanctuaries managed by the National Oceanic and Atmospheric Administration (NOAA), a federal agency focused on the condition of the oceans and the atmosphere. The 842-square-mile sanctuary is located in Massachusetts Bay between Cape Cod and Cape Ann, twenty-five miles east of Boston. The sanctuary is a repository for numerous maritime heritage resources including the shipwrecked *Portland*. The wreck of the schooner *Addie E. Snow* was found resting on the bottom a quarter mile away.

After the initial discovery of the *Portland* wreck, scientists, in cooperation with the National Undersea Research Center at the University of Connecticut, used a remotely operated vehicle (ROV) to gather images of the wreck to confirm its authenticity. According to the Stellwagen Bank National Marine Sanctuary, "Since there were no survivors from the Portland the archaeological remains are the only remnants left that might unveil why it sank. The intact remains of the steamship offer historians and archaeologists the chance to examine a virtual time capsule of information from that fateful night in November 1898."

Images show that the *Portland* rested upright on the ocean's floor with its wooden hull intact from its keel up to the main deck level. The ship's whole superstructure is gone, with only the walking beam and associated engine machinery bulging up some twenty-five feet from deck level. The ROV images showed that the *Portland*'s rudder was still in place and "the deck planking on the main deck, the deck beams, frames, outer hull planking, and copper alloy sheathing are all nearly intact."

> Marine scientists announced today they had found the paddle steamship *Portland*, known as the "*Titanic* of New England," which sank in a wintry gale. . . . Researchers had spent decades looking for the 291-foot vessel. Sonar images and digital video from remotely operated vehicles confirmed the find this week. The ship was found in the Gerry E. Studds-Stellwagen Bank National Marine Sanctuary, an area the size of Rhode Island between Cape Ann and Cape Cod. "This discovery closes the chapter on one of the greatest maritime disasters in New England," said Benjamin D. Cowie-Haskell, the National Oceanic and Atmospheric Administration's primary investigator of the Portland expedition. (Pamela Ferdin, *Washington Post*, August 30, 2002)

The *Portland* wreck was also seen to be ensnarled with many fishing nets. Most of the nets are snarled on the highest points of the wreck. Other nets surround the bow and stern. The amount of nets on the wreck limited the ability of the ROV to provide a detailed analysis of the entire site.

Based on the 2002 investigation, it was reported that the most intact portion of the *Portland* was its 291-foot-long wooden hull. According to information based on the investigation,

> Its keel, frames, and portions of outer hull planking are made of white oak and are in an excellent state of preservation. . . . In some

areas the deck planking has eroded away exposing the massive deck beams. . . . The paddle wheel is almost entirely complete. Machinery on the site include portions of both paddle wheels, the paddle wheel flanges, the paddle shaft, walking beam, and wooden A frame. . . . Both 23-foot long boilers are present and in their original location based upon the position of the boiler uptakes. . . . The steering gear . . . is still affixed to the deck in its original position; however the anchor windlass has fallen through the deck into the chain locker. . . . Overall, the appearance of the wreck is that of a vessel that sank intact. (Deborah Marx, "The Steamship Portland: Update on Stellwagen Bank National Marine Sanctuary's, Investigation of the Site," *Historic Naval Ships Association*, 2007)

Steamship Portland 2003

Sept. 13–Sept. 18, 2003

Researchers from NOAA's Stellwagen Bank National Marine Sanctuary, along with the NOAA-UConn team and filmmakers from The Science Channel, returned to the wreck of the famed 19th-century steamship from Sept. 13–Sept. 18, 2003. Kicking off the expedition to peer into the vessel's past and plan for its future, the team conducted the first surveys of the *Portland* since its location was confirmed in August 2002 within NOAA's Stellwagen Bank National Marine Sanctuary off the Massachusetts coast. (*Ocean Explorer*, NOAA, 2003)

23: Further Investigations

In October 2008 divers again descended to the ocean's depths to investigate the *Portland* wreck. Five Massachusetts men, including Bob Foster of Needham, David Faye of Cambridge, Don Morse of Beverly, Paul Blanchette of Dracut, and Slav Mlch of East Boston, became the first human divers to reach the *Portland* since its sinking.

It took them seven different tries to reach the vessel. Because the location of the wreck has been kept secret to discourage salvagers, the divers had to independently verify the location. They had no sponsors and paid out of pocket for their dives. The diving equipment cost between $10,000 and $50,000 per person, according to David Faye.

"The very first thing we saw was the walking beam, part of the steam engine, sticking up fifteen feet above the wreck," Blanchette said. "Swimming past it on the first dive was impressive." The divers said they found evidence of those who went down with the ship during the storm, but uncovered no human remains.

> Smaller cultural artifacts lie scattered inside and outside the hull and remind us of the passengers and crew that lost their lives when the *Portland* sank. Stacked plates and cups lie exposed on the main deck in the kitchen, while other pieces of dishware, an electric light, a toilet, and a stack of glass window panes have fallen to the seafloor just outside of the hull. A single large mug or cup also rests amongst the tangled steam piping near the steering gear towards the bow. . . . The hull settled to the bottom without much velocity leaving large machinery structures and fragile artifacts intact and near to their original positions. (Deborah Marx, "The Steamship Portland: Update on Stellwagen Bank National Marine Sanctuary's, Investigation of the Site," *Historic Naval Ships Association*, 2007)

Their dives were so deep that some of the underwater lights blew out. The divers used a combination of helium and oxygen in their tanks and it reportedly took up to four hours for them to surface from the depth to avoid decompression sickness. According to Blanchette, who was the leader of the group of divers, the surface water temperature was 68 degrees, making it comfortable to decompress when coming up. Below one hundred feet the temperature was 43 degrees, he said. The divers could only spend ten to fifteen minutes exploring the wreckage site before returning to the surface.

"You have to give the wreck its respect," David Faye said. "It's deep and it's dangerous, and you have to be at the top of your game."

19th Century Steamship Wreckage Reached

> Five Massachusetts men became the first divers to reach the wreck of a 19th-century steamship that sank in one of the most destructive storms in New England history. . . . "I immediately thought of these people—how horrible it must've been," David Faye, a lawyer in Cambridge, Mass. said. "They had no communication with shore. They had no idea where they were. The storm was pushing them out to sea." (*CBS News*, October 7, 2008)

24: Why?

No one will ever know why Captain Blanchard decided to make the return trip to Boston on the evening of Friday, November 26, 1898. Clearly he had checked the weather reports. It may just be that he was confident that, regardless of the forecast, he could stay ahead of the storm and make it safely back to Portland Harbor. Nor can it be known why he didn't turn the *Portland* around and make for safe harbor when it became clear to him that he was sailing straight into a furious winter storm. And whether Portland Steamship Company officials had contacted Blanchard by telephone and explicitly told him to remain docked at India Wharf in Boston, as they claimed, or if he had been ordered to set for home that evening, no one will ever know. All that remains is speculation.

The loss of the *Portland* was New England's worst maritime disaster, with the greatest loss of life prior to 1900. It remains a tragedy of epic proportions and the memory of it still conjures up the enduring struggle of Man against the forces of Nature.

Acknowledgments

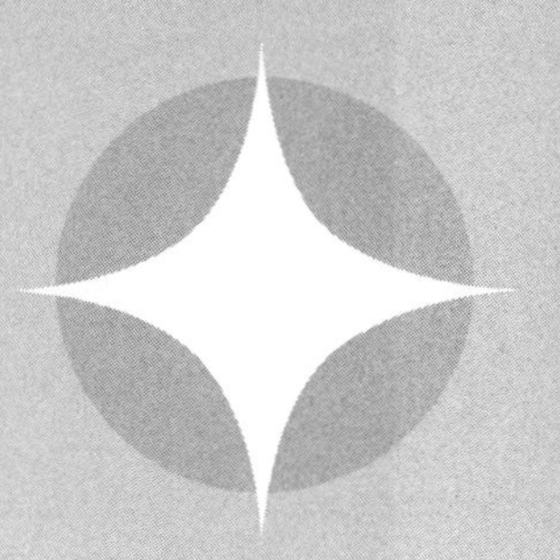

I wish to thank Lyons Press for their continued support and my extraordinary editor there, Tom McCarthy, who makes my books better than I ever dreamed they could be. I wish to thank my lovely wife, Julia, my fiercest critic and staunchest supporter. I would also like to thank my friend and fellow author, Dr. Michael J. Vieira, who assisted me with the many technological hurdles this book presented. As I have often told people, I am, "a workhorse," not a "show horse." I am a storyteller, not a historian. I am a chronicler of events and people. Lyons Press, Tom McCarthy, Julia and Mike Vieira have made it easier to be all these things and they have collectively made my labors worthwhile. Thank you.

Appendices

List of *Portland Crew*, November 1898

Source: *Prologue Magazine* 38, no. 4 (Winter 2006)

Note: "This crew list for the final voyage of the *Portland* was reconstructed primarily with information obtained from the wage papers and claims filed with the U.S. circuit court. Information about race was taken from the U.S. population census of 1900 and the Canadian provincial censuses of New Brunswick and Nova Scotia of 1901. When census records were not available, that column was left blank."

Allen, Henry George	Porter	Harrisonburg, VA
Barron, Matthew	Deckhand	Placentia, Newfoundland
Berry, Mrs. Marjorie A.	Stewardess	Yarmouth, ME
Blake, Rodney S.	Watchman	Brooklin, ME
Blanchard, Hollis H.	Master	Belfast, ME
Bruse, Denis	Deckhand	N. Roms Island, Newfoundland
Carter, Allen	Fireman	Kouchibouguae, New Brunswick
Cash, William H.	Saloon	Wilmington, NC

Collins, Peter L.	Deckhand	St. John, New Brunswick
Cropley, George H.	Deckhand	
Crozier, John	Deckhand	Willow Grove, St. John, New Brunswick
Daley. John	Deckboy	Cork, Ireland
Dauphinee, Everett	Deckhand	Chester, Nova Scotia
Davidson, James	Deckhand	DeBert, Nova Scotia
Dillon, John A.	Oiler	Eastport, ME
Doughty, William	Fireman	Eastport, ME
Dunn, William	Saloon	
Dyer, Ansel Lewis	Qtrmaster & 2nd Mate	Portland, ME
Foreman, Lee Steam	Tableman	South Hampton, VA
Gately, John K.	Fireman	Portland, ME
Gatling, Alexander	Saloon	Elizabeth City, VA
Graham, George H.	Cabin Man	Burlington, NJ
Graham, Maurice	Deckhand	Simonds, New Brunswick
Harris, Mrs. Carrie E.M.	Stewardess	St. Mary's Bay, Nova Scotia
Hartley, Richard	Deckhand	Northeast, Newfoundland
Hemenway, William A.	Cabin Man	Worcester, MA
Heuston, Francis E.	Chief Steward	
Howard, Stephen	Cook	St. John, New Brunswick
Ingraham, Frederick A.	Purser	
Johnson, Charles H.	Saloon	
Johnson, Arthur A.	Saloon Watchman	Moose River, Nova Scotia
Jones, John	Cook	Frederick, MD

Latimer, William E.	Head Saloon Man	St. Croix, West Indies
Leighton, Franklin	Electrician	Falmouth, ME
MacKey, John	First Mate	
Matthews, Alonzo V.	Steward	
McGillivray, George	Deckhand	St. John, New Brunswick
McNeil, James J.	Oiler	St. John, New Brunswick
Merrill, Thomas B.	Chief Engineer	Norwalk, CT
Merrill, Charles L.	2nd Asst. Engineer	Westbrook, ME
Merriman, Hugh	Fireman	Harpswell, ME
Minott, Michael	Saloon	Port Antonio, Jamaica
Moore, Horace C.	Clerk	Portland, ME
Mundrucu, Theodore M.C.	Saloon Watchman	Pernambuco, Brazil
Nelson, Lewis Martin	Pilot	Norway
Norton, George A.	Deckhand	Lubec, ME
O'Brien, Cornelius	Deckhand	
Oxley, Ernest	Pantryman	Port Antonio, Jamaica
Patterson, Frank A.	2nd Mate	Belfast, ME
Pennell, Thomas H.	Fireman	Portland, ME
Pinna, Roland J.	Cabin Man	Cape Verde Islands
Reed, Griffin S.	Forward Cabin Watch	Portland, ME
Robichau, Winthrop P.	Baggage Master	New Brunswick
Rollinson, Harry C.	Fireman	Eastport, ME

Sewall, Thomas	Watchman	Westport, ME
Sloan, Arthur	Deckhand	Willow Grove, St. John, New Brunswick
Smith, Fred	Deckhand	Deer Isle, ME
Smith, Samuel Henry	Hall Man	Lynchburg, VA
Stanley, James	Deckhand	Brooklin, ME
Thompson, William G.A.	Cabin Man	
Walton, John Tuck	1st Asst. Engineer	Cape Elizabeth, ME
Whitten, John C.	Watchman	
Williams, James	After Cabin Watch	Bermuda
Wills, Fred A.	Cook	St. Martin, West Indies

List of Known *Portland* Passengers, November 1898

Source: Gerry E. Studds, Stellwagen Bank National Marine Sanctuary

Note: "The following information is based upon various Maine and Massachusetts newspapers, documents in historical societies and maritime museums around New England, court papers held by the National Archives Northeast Region in Waltham, MA, and the book *Four Short Blasts: The Gale of 1898 and the Loss of the Steamer Portland* by Peter Dow Bachelder and Mason Philip Smith."

Allen, Frederick H., Portland, ME
Atamian, George, Portland, ME
Baker, Beulah Mosley, Portland, ME
Beardsworth , William, Boothbay, ME
Bemis, Cora V., Portland, ME
Bemis, Walter L., Portland, ME
Bonney, Alonzo G., Portland, ME

Briggs, –, Portland, ME
Brown, Fred A., Portland, ME
Buckminster, Joseph, Yarmouth, ME
Carroll (Mrs.) J. A., Portland, ME
Chase, Philip A., Portland, ME
Chase William L., Portland, ME
Chickering, Abbie C., Portland, ME
Clark, Albert, Portland, ME
Cobb, Emily L., Boston, MA
Cohen, Solomon, Portland, ME
Cole, George W., Weymouth, MA
Collins, Elizabeth M. A., Weymouth, MA
Cottreau, Charles C., Worcester, MA
Curtis, Elixabeth, Worcester, MA
Daly, Jerry, Woodfords, ME
Delaney, George E., Woodfords, ME
Dennis, (Mrs.) Ezekiel, Woodfords, ME
Doherty, John, Chelsea, MA
Dunbar Albert B., Chelsea, MA
Dwyer, W., East Boston, MA
Dyer, Charles G., Easton, ME
Dyer, N. Easton, ME
Edmunds, Jennie G., East Boston, MA
Edwards, Lawrence, S. Portland, ME
Flower, Charles S. Portland, ME
Flower, James W. S. Portland, ME
Foden, Rachael, Portland, ME
Foden, Robert, Portland, ME
Fowler, Frederick, Portland, ME
Freeman, Dudley E., Woodfords, ME
Frye, Isaiah, Portland, ME
Frye, Ruth, Portland, ME
Galley, John G., Brookline, MA
Gatchell, Dennis O., Portland, ME
Gately, Ellen D., Portland, ME
Gibson, George P., Portland, ME

Goggin, (Mr.) A. A., Portland, ME
Gosselin, Alphonso, Portland, ME
Hanley, Matthew, Portland, ME
Hanson, William, S. Portland, ME
Heald, Rowena M., Gorham, ME
Hersom, Arthur F., Cumberland, ME
Hersom, (Mrs.) Arthur F., Deering, ME
Holmes, Sophie B., Marlboro, ME
Hooper, Carl, Lewiston, ME
Hooper, Oren, Marlboro, ME
Hoyt, Cynthia J. Providence, RI
Ingraham, Madge, Auburndale, MA
Jackson, Lillian, Auburn, ME
Jackson, Malcolm, Auburn, ME
Jackson, Perry, Ligonia, ME
Kelley, Susan A., Somerville, MA
Kennedy, William, Boston, MA
Kenniston, George B., S. Portland, ME
Kinyon, Florence M., Portland, ME
Kirby, Timothy, Portland, ME
Langthorne, Helen M., Lowell, MA
Leighton, Diana, Manchester, NH
Leighton, Ora L., Gorham, ME
Lord, Hattie A., Boston, MA
Mann, John G., Malden, MA
Matthews, Albert, S. Portland, ME
Matthews, (Mrs.) Albert, Everett, MA
Matthews, –, St. John, Nova Scotia
McGilvery, D. W., Roxbury, ME
McGowen, Lilla, Portland, ME
McKinney (Mrs.) –, Lowell, MA
McMullen, Jennie G., Portland, ME
Metcalf, Lewis J., Portland, ME
Mitchell, Cornelia N., Gloucester, MA
Morong, Faith, Cincinnati, OH
Mosher, William J., Cincinnati, OH
Munn, William F., Portland, ME

Murphy, John H., Westbrook, ME
Murphy, John J., Auburn, ME
O'Connell, John, Auburn, ME
Piche, Jules, Suffield, CT
Plympton, Emma L., Boston, MA
Pratt, Amy, Portland, ME
Pratt, Hannah, Portland, ME
Prescott, George L., Portland, ME
Proctor, Warren S., Boston, MA
Revenal, Theodore, Auburn, ME
Reynolds, Alice, Bangor, ME
Richardson, Frank F., Albany, NY
Roche, William H., Montreal, Canada
Roddy, James, Bath, ME
Rounds, Agnes, Bath, ME
Safford, Miranda, –
Schmidt, Jes, J., Montreal, Canada
Schmidt, Jessine, Auburn, ME
Schmidt, Jorgen J., – , ME
Schmidt, Anton, Scarboro, –
Sherwood, Frederick R., –
Silverstaine, Harry, Portland, ME
Silverstaine, Louis, Boston, MA
Small, Myrton L., Portland, ME
Smith, Harry, Cambridge, MA
Stanley, James, –
Sullivan (Mrs.) John, Brunswick, ME
Swift, Ella M., Bath, ME
Sykes, Maud, Portland, ME
Tetrow, Annie, Hanson, MA
Thompson, Charles H., Deering, ME
Thompson, Gladys M., Rockland, ME
Thompson, Susan E., Hartford, CT
Tibbets, Charles A., Berlin, NH
Timmons, Elmira B., Yarmouth, Nova Scotia
Tinkham, Charles C. Reading, MA
Totten, Eva M., Boston, MA

Tucker, Alice, Bath, ME
Tupper, James H., Portland, ME
Turner, Augustus R., Philadelphia, PA
Turner (Mrs.) Augustus, Yarmouth, Nova Scotia
Twombly, Ella, Portland, ME
VanGuysling, C. E., Portland, ME
Welch, Mary, Brewer, ME
Wheeler, Eunice A., Portland, ME
White, Horace, Portland, ME
Wiggin, Charles, Portland, ME
Wildes, Alonzo F., Portland, ME
Wilson, Frank, I Lisbon, ME
Young, Henry, Windsor, Nova Scotia

List of Vessels Lost or Damaged in the Portland Gale, November 1898

Source: U.S. Life-Saving Service Annual Reports

A.B. Nickerson (schooner), Cape Cod
A.B. Nickerson (steamer), Cape Cod
Abel E. Babcock, Boston Bay
Abby K. Bentley, Vineyard Sound
Addie E. Snow, Cape Cod
Addie Sawyer, Vineyard Sound
Adelaide T. Hither, Plain, NY
Africa, Portland, ME
Agnes, Cape Cod
Agnes May, Cape Ann
Agnes Smith, Pt. Judith, RI
Albert H. Harding, Boston Bay
Albert L. Butler, Cape Cod
Alida, Islesboro, ME
Aloha, Block Island, RI
Anna Pitcher, Block Island, RI
Anna W. Barker, Southern Island, ME

Annie Lee, Cape Ann
Arabell, Block Island, RI
B.R. Woodside, Boston Bay
Barge (unknown), Boston Bay
Barge (unknown), Boston Bay
Barge (unknown), Boston Bay
Barge (unknown), Boston Bay
Barge (unknown), Boston Bay
Barge (unknown), Cape Ann
Barge (unknown), Cape Ann
Barge No. 4, Boston Bay
Beaver, Vineyard Sound
Bertha A. Gross, Cape Ann
Bertha E. Glover, Vineyard Sound
Brunhilde, Point of Woods, NY
Byssus, Vineyard Sound
C.A. White, Boston Bay
C.B. Kennard, Boston Bay
Calvin F. Baker, Boston Bay
Canaria, Vineyard Sound
Carita, Vineyard Sound
Carrie C. Miles, Portland, ME
Cassina, Block Island, RI
Catboat (unknown), Short Beach, NY
Cathie C. Berry, Vineyard Sound
Champion, Cape Cod
Charles E. Raymond, Vineyard Sound
Charles E. Schmidt, Cape Ann
Charles J. Willard, Quoddy Head, ME
Chillion, Cape Ann
Chiswick, Boston Bay
Clara Sayward, Cape Cod
Clara P. Sewall, Boston Bay
Columbia, Boston Bay
Consolidated Barge No. 1, Boston Bay
D.T. Pachin, Cape Ann
David Boone, Cape Cod

Daniel L. Tenney, Boston Bay
David Faust, Nantucket
Delaware, Boston Bay
E.G. Willard, Vineyard Sound
E.J. Hamilton, Vineyard Sound
Earl, Edith, Cuttyhunk
Edgar J. Foster, Boston Bay
Edith McIntire, Vineyard Sound
Edna & Etta, Great Egg, NJ
Edward H. Smeed, Block Island, RI
Ella F. Crowell, Boston Bay
Ella Frances, Cape Cod
Ellen Jones, Cape Cod
Ellis P. Rogers, Cape Ann
Elmer Randall, Boston Bay
Emma (wreckage), Boston Bay
Ethel F. Merriam, Cape Cod
Evelyn, Cape Ann
F.H. Smith, Cape Cod
F.R. Walker, Cape Cod
Fairfax, Cuttyhunk
Fairfax, Vineyard Sound
Falcon, Vineyard Sound
Fannie Hall, Portsmouth, NH
Fannie May, Rockland, ME
Flying Cloud, Cape Ann
Forest Maid, Portsmouth, NH
Fred A. Emerson, Boston Bay
Friend, Cuttyhunk
Fritz Oaks, Boston Bay
G.M. Hopkins, Boston Bay
G.W. Danielson, Block Island
Gatherer, Cape Ann
George A. Chaffee, Cape Ann
George H. Miles, Vineyard Sound
Georgietta, Spruce Head, ME

Grace, Cape Cod
Gracie, Cape Cod
Hattie A. Butler, Vineyard Sound
Henry R. Tilton, Boston Bay
Hume, Boston Bay
Hurricane, Rockland, ME
Ida, Boston Bay
Ida G. Broere, Lone Hill, NY
Idella Small, Davis Neck, Mass.
Inez Hatch, Cape Cod
Institution (launch), Boston Bay
Ira and Abbie, Block Island, RI
Ira Kilburn, Portsmouth, NH
Isaac Colline (schooner), Cape Cod
Isaac Collins (schooner), Cape Cod
Island City, Vineyard Sound
Ivy Bell Jerry, Race Point, NH
J.C. Mahoney, Cape Ann
J.M. Eaton, Cape Ann
James A. Brown, Vineyard Sound
James Ponder, Vineyard Sound
James Webster, Boston Bay
John Harvey, Point Judith, RI
John J. Hill, Boston Bay
John S. Ames, Boston Bay
Jordon L. Mott, Cape Cod
Juanita, Boston Bay
King Phillip, Cape Cod
Knott V. Martin, Cape Ann
Leander V. Beebe, Boston Bay
Leora M. Thurlow, Vineyard Sound
Lester A. Lewis, Cape Cod
Lexington, Block Island, RI
Lillian, Portland, ME
Lizzie Dyas, Boston Bay
Lucy A. Nichols, Boston Bay

Lucy Bell, Boston Bay
Lucy Hammond, Vineyard Sound
Luther Eldridge, Nantucket
M.E. Eldridge, Vineyard Sound
Marion Draper, Vineyard Sound
Mary Cabral, Cape Cod
Mary Emerson, Boston Bay
Mascot, Cuttyhunk
Mertis H. Perry, Boston Bay
Michael Henry, Cape Cod
Mildred and Blanche, Cape Cod
Milo, Boston Bay
Multnoman, Portsmouth, NH
Nautilus, Cape Cod
Nellie B., Block Island
Nellie Doe, Vineyard Sound
Nellie M. Slade, Vineyard Sound
Neptune, Portland, ME
Neverbuge, Cuttyhunk
Newburg, Vineyard Sound
Newell B. Hawes, Plum Island, MA
Ohio, Boston Bay
Papetta, Vineyard Sound
Pentagoet, Cape Cod
Percy, Block Island, RI
Phantom, Boston Bay
Philomena Manta, Cape Cod
Pluscullombonum, Boston Bay
Portland, off Cape Cod
Powder vessel (unknown), Boston Bay
Queen of the West, Fletcher's Neck, ME
Quesay, Vineyard Sound
Rebecca W. Huddell, Vineyard Sound
Reliance, Point of Woods, NY
Rendo, Portland, ME
Rienzi, Cape Ann

Ringleader, Portsmouth, NH
Robert A. Kenner, Boston Bay
Rose Brothers, Block Island, RI
Rosie Cobral, Boston Bay
S.F. Mayer, Rockland, ME
Sadie Wilcutt, Vineyard Sound
Sarah, Cape Ann
School Girl, Cape Cod
Schooner (unknown), Boston Bay
Schooner (unknown), Boston Bay
Schooner (unknown), Boston Bay
Schooner (unknown), Boston Bay
Schooner (unknown), Cape Cod
Schooner (unknown), Cape Cod
Secret, Cuttyhunk
Seraphine, Boston Bay
Silver Spray, Portland, ME
Sloop (unknown), Boston Bay
Sloop (unknown), Boston Bay
Sloop (unknown), White Head, ME
Sport, Cuttyhunk
Startle, Boston Bay
Stone Sloop (unknown), Boston Bay
Stone Sloop (unknown), Boston Bay
Stranger, Block Island, RI
Sylvester Whalen, Cape Cod
T.W. Cooper, Portsmouth, NH
Tamaqua, Boston Bay
Thomas B. Reed, Cape Cod
Two Sisters, Portsmouth, NH
Two-Forty, Boston Bay
Union, Boston Bay
Unique, Cape Cod
Unknown vessel, Boston Bay
Valetta, Vineyard Sound
Valkyrie, Block Island, RI

Verona (wreckage), Boston Bay
Vigilant, Cape Cod
Virgin Rocks, Boston Bay
Virginia, Boston Bay
W.H. DeWitt, Cape Ann
W.H.Y. Hackett, Portsmouth, NH
Watchman, Boston Bay
Wild Rose, Cranberry Isles, ME
William Leggett, Cape Ann
William M. Wilson, Washapreague, VA
William Todd, Vineyard Sound
Wilson and Willard, Cape Ann
Winnie Lawry, Vineyard Sound
Wooddruff, Northport, ME

Bibliography

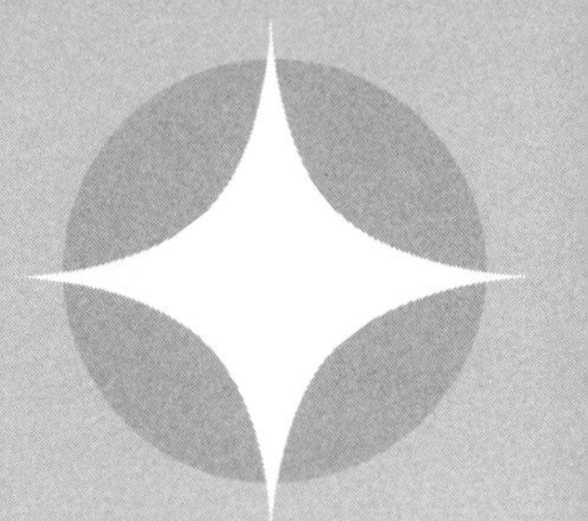

American Citizen, August 17, 1807.

Annual Report of the United States Life-Saving Service, United States Life-Saving Service. Washington, DC: U.S. Government Printing Office, 1914.

Bachelder, Peter Dow, and Mason Philip Smith. *Four Short Blasts: The Gale of 1898 and the Loss of the Steamer Portland*. Portland, ME: The Provincial Press, 2003.

Bacon, George Fox. *Portland, Maine, Its Representative Businessmen and Its Points of Interest*. Newark, NJ: Glenwood, 1891.

Bacon, Kezia. "The Storm That Changed the River's Flow." *Mariner Newspapers*, 2008.

Barbo, Theresa M. *True Accounts of Yankee Ingenuity and Grit from the Cape Cod Voice*. Stroud, UK: The History Press, 2007.

Baxter, Sylvester. "The Great November Storm of 1898." *Scribner's Magazine* 26, no. 5 (November, 1899).

Berman, Bruce D. *Encyclopedia of American Shipwrecks*. N.p.: The Mariners Press. Inc., 1972.

Bipasha, Ray. "Steamship Wrecked in 1898 Found: The *Portland* sank off Mass. in Hurricane." *Bangor Daily News,* August 30, 2002.

Blackington, Alton H. *Yankee Yarns*. New York: Dodd, Mead and Co., 1954.

Bolster, W. Jeffrey. *Black Jacks: African American Seamen in the Age of Sail*. Cambridge, MA: Harvard University Press, 1998.

Borrelli, Peter. *Stellwagen: The Making and Unmaking of a National Marine Sanctuary*. Lebanon, NH: University Press of New England, 2009.

The Boston Globe, November 29, 1898.

———, December 7, 1898.

Boston Herald, November 30, 1898.

———, November 30, 1898.
———, December 2, 1898.
———, December 3, 1898.
Boston Journal, November 28, 1898
Bouchard, Kelley. "Nearly 200 Victims of the Sinking of SS *Portland* Are Remembered," *Portland Press Herald*, December 1, 2017.
Brandt, Anthony, ed. *The Tragic History of the Sea: Shipwrecks from the Bible to Titanic.* Washington, DC: National Geographic Books, 2007.
Brooklyn Daily Eagle, September 2, 1807.
———, Nov. 30 1898.
Cann, Donald J., and John J. Galluzzo. *The Coast Guard in Massachusetts.* Mt. Pleasant, SC: Arcadia Publishing, 2011.
"Cape Cod's Reputation as a Graveyard." *Cape Cod Times*, January 4, 2011.
Chapter 271, Twenty-Fifth Legislature of the State of Maine, 1845. http://legislature.maine.gov/lawlibrary/legislative-history/9211
Chase, Virginia. "Shipwreck: The Portland Disaster." *Downeast* 4, no. 4 (January 1958).
Cahill, Robert Ellis. *Finding New England's Shipwrecks and Treasure.* Danvers, MA: Old Saltbox Publishing, 1984.
Chatterton, Edward Keble. *Steamships and Their Story.* London: Cassell and Company, 1910.
Claflin, James. *Lighthouses and Life Saving along Cape Cod.* Charleston, SC: Arcadia Publishing, 2014.
Congressional Record, V. 145, Pt. 14, August 4, 1999 to August 5, 1999. Washington, DC: Government Printing Office.
"Crushed by Angry Waves, Echoes of the Great Gale." *New York Times*, December 1, 1898.
Dalton, John Wilfred. *Life Savers of Cape Cod.* Boston: The Barta Press Printers, 1902.
Deeson, A. F. L. *An Illustrated History of Steamships.* Buckinghamshire, UK: Spurbooks, 1976.
Duncan, Archibald. *Adventures and Perils: Being Extracts from the 100-years-old Mariner's Chronicle and Other Sources Descriptive of Shipwrecks and Adventures at Sea.* London: Michael Joseph, 1936.
Elgersman Lee, Maureen. *Black Bangor: African Americans in a Maine Community.* Lebanon: University of New Hampshire Press, 2005.
Ferdin, Pamela. *Washington Post.* August 30, 2002.
Finch, Robert. *A Place Apart: A Cape Cod Reader.* Woodstock, VT: The Countryman Press, 2009.

Fletcher, R. A. *Steamships: The Story of Their Development to the Present Day*. New York: J.B. Lippincott Company, 1991.

Freitas, Fred, and Dave Ball. *Warnings Ignored! The Story of the Portland Gale–November 1898*. Scituate, MA: Converpage, 1995.

Gentleman's Magazine, December, 1809.

Gratwick, Harry. *Historic Shipwrecks of Penobscot Bay*. Charleston, SC: History Press, 2014.

Hall, Thomas. *Shipwrecks of Massachusetts Bay*. Charleston, SC: History Press, 2012.

Handbook of New England. Boston: P.E. Sargent, 1921.

Harper's Magazine, vol. 123, 1911.

Heit, Judi ."The Portland Gale," 2010. http://portlandgale.blogspot.com/.

Hickey, Walter V. "The Final Voyage of the *Portland*." *Prologue Magazine*, 2006

Hughes, Lisa. "Steamship Portland Remembered as the *Titanic* of New England." *WBZ-TV*, April 11, 2012. https://boston.cbslocal.com/2012/04/11/steamship-portland-remembered-as-the-titanic-of-new-england/.

Hunter, Louis. *Steamboat on the Western Rivers: An Economic and Technological History*. New York: Dover Maritime, 1949.

Johnson, William, and Kin Knox Beckius. *The New England Coast*. Minneapolis, MN: Voyager Press, 2008.

Kittredge, Henry C. *Mooncussers of Cape Cod*. New York: The Riverside Press. 1937.

Laird, Neil, writer, producer. *Ghost Ship of New England: Deep Sea Detectives*. DVD. The History Channel, 2008.

Lawrence, Mathew. "Lost and Found: The Search for the Portland." *Sea History*, no. 107 (Spring/Summer 2004).

Mathew, Lawrence, Deborah Marx, and John Galluzzo. *Shipwrecks of Stellwagen Bank: Disasters in New England's National Marine Sanctuary*. Charleston, SC: History Press, 2015.

Library of Congress. "Weather Facts," 1949.

"Losses of Marine Underwriters." *Boston Herald*, December 2, 1898.

Marx, Deborah. "Forbidden to Sail: The Steamship *Portland*, 1890–1898." *Sea History*, no. 107 (Spring/Summer 2004).

———. "The Steamship *Portland*: Update on Stellwagen Bank National Marine Sanctuary's, Investigation of the Site." *Historic Naval Ships Association*, 2007. https://archive.hnsa.org/conf2004/papers/marx.htm.

"Massachusetts: Last Voyage," *Time*, December 6, 1948.

McLean, Duncan. Boston Daily Atlas, 1851.

Melton, Mary. *Lost with all Hands: A Family Forever Changed, the Portland Gale of 1898*. Penobscot, ME: Penobscot Press, 1998.

Milmore, Art. *And the Sea Shall Not Have Them All.* N.p.: CreateSpace Independent Publishing, 2017.

New York Times, November 29, 1898.

———. November 30, 1898.

———. December 1, 1898.

———. December 27, 1898.

———. March 12, 1899.

Omaha Bee. June 1898.

Park, Charles E. "The Development of the Clipper Ship," *The American Antiquarian Society*, 1929.

Perley, Sidney. *Historic Storms of New England.* Salem, MA: Salem Press, 1891.

The Quarterly Review of London. 1830.

Quinn, William P. *Shipwrecks Around Maine.* Orleans, MA: The Lower Cape Publishing Co., 1983.

"Remembering the Sinking of the Portland." *New England Boating*, April 18, 2012.

Richardson, John. "Ship of Doom." *Blethen Maine Newspapers*, November 22, 1998.

Richardson, John M., and Thomas Harrison Eames. *Steamboat Lore of the Penobscot.* Kennebec, ME: Kennebec Journal Print Shop. 1945.

Ridgely-Nevitt, Cedric. *American Steamships on the Atlantic.* Wilmington: University of Delaware Press, 1981.

Rousmaniere, John. *After the Storm: True Stories of Disaster and Recovery at Sea.* New York: International Marine/McGraw-Hill, 2002.

Sacramento Daily Union 96, no. 102 (December 1, 1898).

Seymour, Tom. *Fishermen's Voice*, October 2014.

Small, Isaac M. *Shipwrecks on Cape Cod: Highlands-Coast of Cape Cod Massachusetts.* Chatham, MA: Old Chatham Press, 1928.

Snow, Edward Rowe. *Marine Mysteries and Dramatic Disasters of New England.* New York: Dodd Mead, 1976.

———. *Great Storms and Famous Shipwrecks of the New England Coast.* Boston: Yankee Publishing Company, 1943.

———. *Storms and Shipwrecks of New England.* Boston: Commonwealth Editions, 2003.

Spotts, Peter. "Scientists Find the '*Titanic* of New England.'" *Christian Science Monitor*, August 30, 2002.

"Steamship *Portland* 2003." *Ocean Explorer*, NOAA, 2003. https://ocean-explorer.noaa.gov/explorations/03portland/welcome.html

Thurston, Robert Henry. *A History of the Growth of the Steam Engine.* New York: Appleton and Co., 1883.

Toppan, Andrew. *Haze Gray & Underway*. http://www.hazegray.org/. 1994–2003.

Varricchio, Louis "The Wreck of the S. S. *Portland*: The *Titanic* of New England." *Sun Community News*, March 27 2012. https://www.suncommunitynews.com/articles/the-sun/wreck-s-s-portland-titanic-new-england/.

Ward, Charlie. *Boston Herald,* December 3, 1898.

Woods, Amy. "Shipwrecks along the New England Coasts." *New England Magazine*, November 1904.

Index

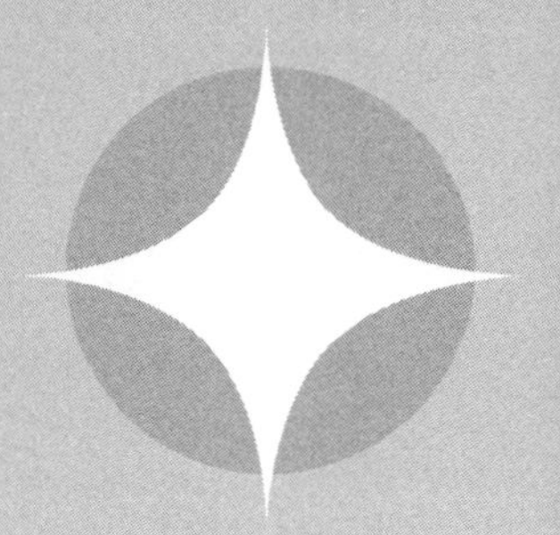

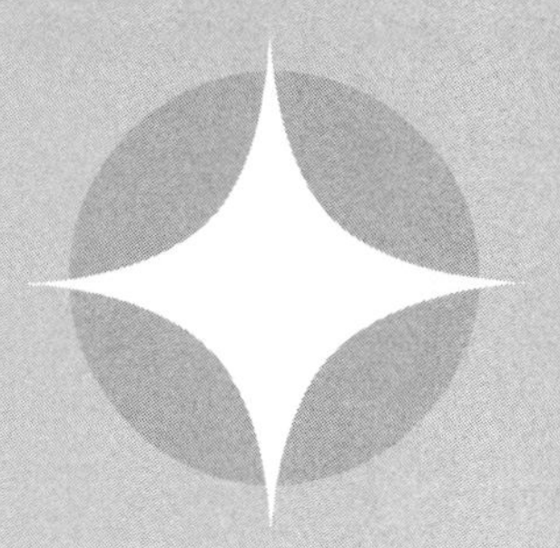

About the Author

J. North Conway is a poet and author of fifteen books. He attended Norwich University military college and served as newspaper reporter and editor for 25 years. He lives in the 200-year-old house of sea captain, John Nichols, in Assonet, Massachusetts (population 3,500), along the banks of the Assonet River and he teaches English at a small community college in southeastern Massachusetts.

Other books by J. North Conway

Soldier Parrott (2019)
New England Rocks (2017)
Outside Providence: Selected Poems (2016)
Attack of the HMS Nimrod: Wareham and the War of 1812 (2014)
Queen of Thieves (2014)
Bag of Bones (2012)
The Big Policeman (2010)
King of Heists (2009)
The Cape Cod Canal: Breaking Through the Bared and Bended Arm (2008)
Head Above Water (2005)
Shipwrecks of New England (2000)
New England Visionaries (1998)
New England Women of Substance (1996)
American Literacy: Fifty Books That Define Our Culture and Ourselves (1994)
From Coup to Nuts: A Revolutionary Cookbook (1987)